MOSBY'S **ANATOMY AND PHYSIOLOGY LABORATORY MANUAL**

MOSBY'S **ANATOMY AND PHYSIOLOGY LABORATORY MANUAL**

KATHLEEN B. SLOAN, B.A.
Honors Biology Instructor, Kingwood High School,
Kingwood, Texas

KATHRYN PERSSON, M.S.
Biology Coordinator, North Harris County College
Houston, Texas

S. KATHLEEN MOORE, B.S.
Research Associate, Optex Biomedical, Inc.,
The Woodlands, Texas

THE C.V. MOSBY COMPANY
St. Louis • Baltimore • Philadelphia • Toronto   1990

Editor: Richard A. Weimer
Developmental editor: Rina Steinhauer
Project manager: Carol Sullivan Wiseman
Production editor: Florence Achenbach
Book designer: Susan E. Lane

Printed in the United States of America

The C. V. Mosby Company
11830 Westline Industrial Drive, St. Louis, Missouri 63146

International Standard Book Number 0-8016-5986-8

GW/VH/VH 9 8 7 6 5 4 3 2 1

PREFACE

The laboratory portion of your study of anatomy and physiology can be informative, challenging, and fun for you, the student. It is a "hands-on" opportunity for you to discover and learn—by yourself or with your classmates.

The new *Mosby's Anatomy and Physiology Laboratory Manual* can provide you with laboratory investigations and demonstrate anatomical and physiological structure and function, yet does not require elaborate facilities, equipment, supplies, or extensive time.

The lab manual organization allows you to develop sequential and progressive knowledge. Many of the initial investigations are incorporated into later investigations so that you can reinforce basic concepts.

Each lab uses an introduction to the subject matter, the objectives of the experiment, and a protocol for the investigation. In this way you will become familiar with the format of scientific research journal literature.

All students should develop the scientific method skills of making an observation, forming a hypothesis, testing the hypothesis, and analyzing the results to form a conclusion. These skills may be enhanced as your teacher creates the environment in an anatomy and physiology laboratory that correlates with the environment in a research laboratory. Therefore I suggest that, with the approval of your teacher, you use a lab notebook to set up each investigative protocol, as well as record the results. As a follow-up to many of the labs, your teacher may ask that a formal lab report be written. Suggested formats for the lab notebook and lab reports are shown in the Appendix.

And now I challenge you to use these lab exercises to learn—but more importantly to enjoy—the fascinating world of anatomy and physiology.

Kathleen B. Sloan

CONTENTS

COLOR PLATES

(appear between pages 56 and 57)

MICROSCOPE

OBJECTIVES

1 Describe proper handling and care of the microscope.
2 Identify and explain the functions of all parts of the microscope.
3 Compute magnification for all objectives.
4 Sketch observations of microscopic organisms.

MATERIALS

binocular stereoscopic
   microscope
compound microscope
prepared slides with the
   letter "e"

Elodea aquatic plant
microscope slide
coverslip
lens paper

KEY TERMS

Aperture
Arm
Base
Binocular stereoscopic
   microscope
Coarse-focus knob
Compound microscope
Condenser
Diaphragm
Fine-focus knob
Magnichanger knob
Objective
Ocular
Parfocal
Stage
Stage clips

Some organisms consist of a single cell, and others consist of many cells. For instance, the human body is estimated as having 100 trillion cells.

The microscope is important in biology, since it enables us to study cells and organisms that are not visible to the human eye. The compound microscope used today is quite sophisticated in comparison to Anton van Leeuwenhoek's simple microscope, through which he first viewed microbes and cells.

The following steps describe proper handling, care, and storage of the microscope:

1. When carrying the microscope, place one hand beneath the base and grasp the arm with the other hand.
2. Avoid bumping or tilting the microscope. The ocular, or eyepiece, can be removed from the body tube of many microscopes and may fall out if the instrument is tilted.
3. The microscope at all times needs to be free from dirt and liquids. A paper towel may be used to wipe any water, stain, or oil from the microscope. Use only lens paper to clean the glass lenses when they are dirty or wet. *Never* place a dirty microscope in the cabinet. Always clean the lenses before and after use.
4. After cleaning the lenses, place the low-power objective in viewing position and turn the coarse adjustment clockwise until the objective is as close to the stage as it will go.
5. The cord should be wrapped properly.
6. Place the plastic cover over the microscope and return it to the storage cabinet.

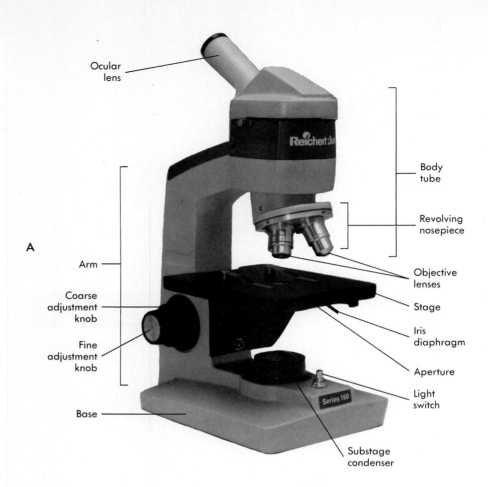

Ocular
lens

Reichert·Ju

Body
tube

Revolving
nosepiece

A

Arm

Objective
lenses

Coarse
adjustment
knob

Stage

Iris
diaphragm

Fine
adjustment
knob

Aperture

Series 160

Light
switch

Base

Substage
condenser

FIG. 1-1   A, Compound light microscope.

The parts and functions of the microscope are as follows (Fig. 1-1, *A* and *B*):

Base   Lowermost part of supporting stand; part of microscope that sits on table

Arm   Part extending up from base

Stage   Platform containing an opening over which slide to be examined is placed

Stage clips   Metal structures that hold slide in place on stage

Condenser (mirror)   Located under stage; focuses light through opening in stage onto object viewed

Iris diaphragm   Shutterlike structure on condenser used to adjust amount of light focused through condenser

Aperture   The circular hole in the stage through which light from the mirror or illuminator passes

Ocular (eyepiece)   Lens-containing structure fitted into the telescoping tube. Remove the ocular from the body tube and note the marking on it. The number indicates the ocular's magnifying power

Objective   Lens-containing structures (usually two or three of them) set in a revolving nosepiece

Coarse-focus adjustment knob   Moves the body tube up or down to bring the image into focus

Fine-focus adjustment knob   When the image comes into view, this adjustment knob is used for finer focus

Magnichanger knob   Allows for changes in magnification without having to refocus *(B)*

The ocular lens is usually 10X. This figure denotes a magnifying power of 10 diameters. The objective lenses range from 10X to 100X. To calculate the total magnification of any object, simply multiply the individual magnifications of the ocular and the objective lenses.

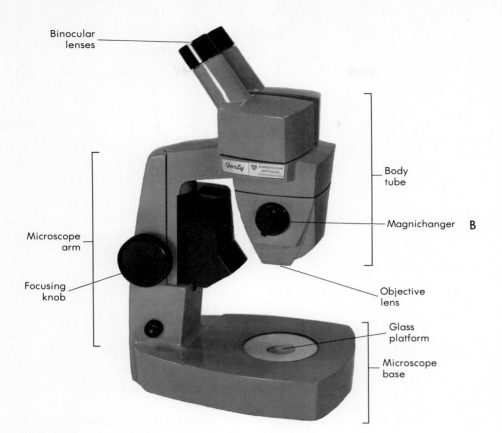

Binocular lenses

Microscope arm

Focusing knob

Body tube

Magnichanger B

Objective lens

Glass platform

Microscope base

FIG. 1-1, cont'd   B, Binocular stereoscopic (dissecting) microscope.

PROCEDURE A   COMPOUND MICROSCOPE OBSERVATIONS
PREPARED SLIDES

1. Obtain a prepared slide with the letter "e."
2. Place slide on the stage and secure with the stage clips.
3. The objectives should always be positioned as far as possible from the stage.
4. Rotate the nosepiece until the low-power objective clicks into position.*
5. Use the coarse adjustment knob to slowly rotate downward (clockwise) toward the slide until an image appears. Then use the fine adjustment knob to make small adjustments to bring the "e" into clear, sharp focus. The illumination may have to be adjusted depending on whether the image is bright or dark.
6. Move the "e" to the left and to the right and make observations.
7. After observing the "e" on low power, carefully swing the high-power objective into position without moving any knobs. Once the image is brought into sharp focus under low power, it will remain in focus when the high-power objective is turned into position (and only fine adjustments are necessary). This is called **parfocal.**
8. Determine the magnification of the letter "e" on both low power and high power.

---

*Always view specimens with low power first.

FIG. 1-2 Wet mount technique.

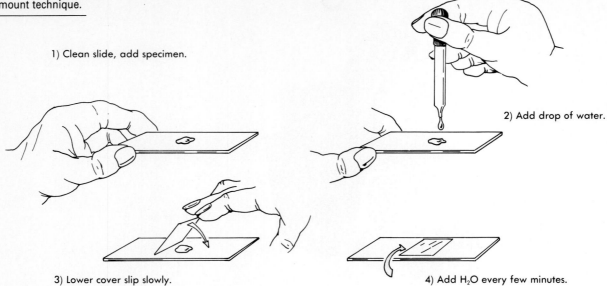

1) Clean slide, add specimen.

2) Add drop of water.

3) Lower cover slip slowly.

4) Add H₂O every few minutes.

PROCEDURE B COMPOUND MICROSCOPE OBSERVATIONS
WET MOUNT (Fig. 1-2)

1. Obtain a single leaf from the aquatic plant **Elodea.** *Elodea* has two layers of cells, which contain chloroplast.
2. Place the leaf and a drop of water on a slide and add a cover slip at a 45-degree angle. Slowly lower the cover slip to the slide. This procedure helps prevent air bubbles from forming on the viewing area.
3. Place the slide on the microscope in the same manner used with the prepared slide, and use the same steps to focus.*
4. Using high power, sketch both layers of cells of the *Elodea.*
5. Clean the microscope after this procedure.

* CAUTION: When using high power, use only the fine adjustment knob to avoid breaking the coverslip.

PROCEDURE C BINOCULAR STEREOSCOPIC MICROSCOPE

1. Obtain various objects to examine under the binocular stereoscopic microscope.

1. Define parfocal and aperture.

2. How do you calculate the magnification for a microscope?

3. List and describe the parts of a microscope.

4. Describe the proper care and handling of a microscope.

5. Describe the procedure for making a wet mount slide.

6. Who was the inventor of the first simple microscope?

*Use all references and materials at your disposal to answer these review questions.

2

WATER MOLECULES

OBJECTIVES

1 Observe how water is a good solvent.
2 Observe how water allows nonpolar molecules to be repelled.
3 Observe how water interacts with hydrophobic/hydrophilic molecules.
4 Observe how water aids in capillary action resulting from adhesion and cohesion.
5 Observe how water maintains a high specific heat.

MATERIALS

250 ml beaker	tap water
balance scale	salt
salad oil	dish soap
glass tube	needle
hot plate	weighing paper
milk	petri dish

Living organisms contain between 50% and 90% water. The human organism is composed of approximately 66% water. Most of life's chemical reactions occur in an aqueous solution.

The water molecule (H_2O) contains one oxygen atom and two hydrogen atoms (Fig. 2-1). In the **covalent bonding** of a water molecule, two hydrogen atoms share electron pairs with two of the four electron pair orbitals in the outer shell of the oxygen atom.

FIG. 2-1 Three diagrams of the polar covalently bound water molecule.

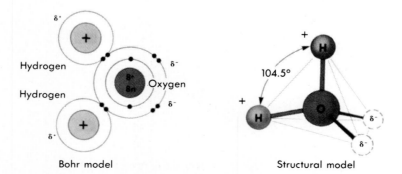

Bohr model　　　　Structural model　　　　Molecular model

The ability of an atom to hold its own electrons and to attract more electrons is called **electronegativity.** When the sharing of electrons involves atoms of different electronegativity, the sharing is unequal, causing a **polar** covalent bond. Since the oxygen atom has a greater electronegativity than hydrogen, the atoms are close together, forming a lopsided molecule. This lopsided formation allows the water molecule to have a positive charge at the hydrogen end and a negative charge at the oxygen end.

The polarity of the water molecule causes the positive end of one water molecule to be attracted to the negative end of another water molecule. The attraction forms a weak **hydrogen bond.**

Due to the polarity of the water molecule and its weak hydrogen bonds, it acts as an excellent **solvent** (a liquid that dissolves another substance). The positive end of the **solute** (the dissolved substance) and the negative end readily interact with other substances to form a **solution** (the combination of solute and solvent).

In table salt (NaCl) the sodium (Na) has one electron in its outer shell, and it is readily given to the chloride (Cl), which contains seven electrons in its outer shell. The single sodium electron is **ionically** bonded to the chloride outer shell. This provides stability for both atoms. The chloride is able to attract the sodium electron because of its strong electronegativity. The sodium now has a positive charge, forming what is called a sodium **ion** (Na^+). The chloride now has a negative charge, forming a chloride ion (Cl^-).

When NaCl is placed in water, it **dissociates.** The negative end of the water molecule is attracted to the positive end of the Na^+. Likewise, the positive end of the water moelcule is attracted to the Cl^-.

When two atoms are covalently bonded in an equitable manner, the molecule is said to be **nonpolar,** or **hydrophobic** (repels water). Salad oil is nonpolar and does not form hydrogen bonds with water. Nonpolar molecules also have little attraction for one another. However, the water molecules have a strong mutual attraction, thereby excluding the oil and pushing the nonpolar molecules together.

Hydrophilic (attracts water) molecules include polar molecules or ions. Some molecules are hydrophilic at one end and hydrophobic at the other. These molecules can be a link between water and fat. Soaps and detergents are nonpolar at one end and charged at the other, so nonpolar fat molecules can be dispersed in dishwater.

The attraction of similar substances, such as one water molecule to another water molecule, is called **cohesion. Adhesion** is the attraction of dissimilar substances such as water and glass. Water will move up a glass tube due to the cohesive and adhesive properties of water. This is known as **capillary action.**

The **specific heat** is the amount of heat required to raise 1 g of a substance 1° C. Because the presence of hydrogen bonds reduces the energy of motion, a broken water molecule will be quickly attracted to another water molecule. Water has a high specific heat and therefore requires a great deal of heat to counteract the attractive forces.

Another property of water is **surface tension.** Because water molecules cannot form hydrogen bonds with air, the strong cohesive surface area of water tends to make the surface behave as an elastic membrane.

PROCEDURE A   WATER AS A SOLVENT

1. Place 0.5 g of salt in a beaker of 50 ml water and stir.
2. Now place 1.5 g of salt in the same beaker of water and stir.
3. Clean beaker.

PROCEDURE B   WATER AND NONPOLAR MOLECULES

1. Add 5 ml salad oil to a beaker of 50 ml water and stir. Describe what happens.
2. Add 5 ml water to a beaker of salad oil and stir. Describe what happens.

PROCEDURE C   DETERGENTS AND SOAPS

1. Add several drops of dish soap to the beaker of water and salad oil to step 1 in procedure B. Stir and describe what happens.

PROCEDURE D   COHESION AND ADHESION

1. Place a glass tube in a beaker of water.
2. Look carefully along the **interface** (where the water meets the air) of the glass tube. Compare the time required for each liquid to boil.

PROCEDURE E   TEMPERATURE STABILITY AND WATER

1. Add some water to one beaker and some milk to another beaker. Place both beakers on a hot plate at maximum temperature. Wait for both to boil. Compare the time required for each liquid to boil.

PROCEDURE F   SURFACE TENSION

1. Fill a petri dish with water.
2. While holding a dry needle at both ends, gently place it on top of the water.
3. If the needle fails to float, repeat the procedure with a dry needle.
4. Once the needle is afloat, add a drop of dish soap as close to the needle as possible. Describe what happens.

1. How is water an important solvent in the lungs?

2. How is water an important solvent in the blood?

3. How is water an important suspending medium in body cells?

4. Explain how water is important in chemical reactions.

5. How is the high specific heat of water important to the body?

6. Discuss how water in the body serves as an excellent lubricant.

7. Explain how phospholipids of the cell membrane are similar to soaps.

8. The cohesive and adhesive properties of water may be observed in which system of the body?

*Use all references and materials at your disposal to answer these review questions.

3

BIOLOGICAL MOLECULES
Carbohydrates

KEY TERMS

Acetyl group
Aldehyde group
Amino group
Benedict's reagent
Carbohydrates
Carboxyl group
Conformation
Covalent bond
Dehydration synthesis
Disaccharide
Disulfide group
Fructose
Functional group
Galactose
Glucose
Hydrogen bond
Hydrolysis
Hydroxyl group
Iodine reagent
Ionic bond
Ketone group
Lipids
Molecular formula
Monosaccharide
Nucleic acids
Phosphate group
Polysaccharide
Proteins
Structural formula
Sucrose

OBJECTIVES

1 **Identify a monosaccharide using a Benedict's test.**
2 **Structurally demonstrate the formation of a disaccharide molecule.**
3 **Identify a starch using an iodine test.**

MATERIALS

hot plate   sucrose solution
250 ml beakers   starch solution
4 clean test tubes   Benedict's reagent
distilled water   iodine reagent
glucose solution   molecule-building models
test tube holder   test tube rack

Biological molecules include four major classes: **carbohydrates, lipids, proteins,** and **nucleic acids.** They may be found alone in our bodies or in varying combinations.

In living organisms the most abundant of the elements are:

Carbon = C   Nitrogen = N
Hydrogen = H   Sulfur = S
Oxygen = O   Phosphorus = P

Particular combinations of these elements have similar properties and are placed in **functional groups.** Biologists use these functional groups to classify biological molecules (Fig. 3-1).

Hydroxyl group   Amino group
Acetyl group   Phosphate group
Carboxyl group   Aldehyde group
Disulfide group   Ketone group

Carbohydrates have the molecular formula $(CH_2O)_n$ in which n means there may be any multiple of CH_2O. Carbohydrates are categorized as either a **monosaccharide, disaccharide,** or **polysaccharide.**

Monosaccharides (simple sugars) are composed of a short chain of carbon atoms (containing two to seven carbons) with a **hydroxyl functional group.** A monosaccharide ring structure will often contain an **aldehyde group** at one end; some monosaccharide ring structures contain a **ketone group.** Examples of these simple sugars are **glucose, fructose,** and **galactose.** The same **molecular formula** (kind and number of atoms) is found in each of these monosaccharides; however, the atoms are arranged differently in their **structural formula.**

A monosaccharide and **Benedict's reagent** will react together in the presence of heat. The free aldehyde group in the sugar (see box on p. 14) is **oxidized** (the addition of an

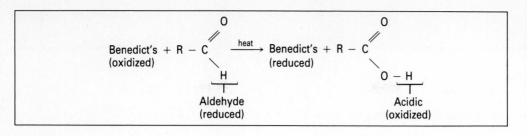

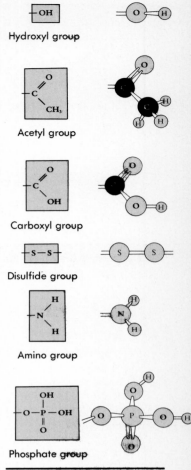

Hydroxyl group

Acetyl group

Carboxyl group

Disulfide group

Amino group

Phosphate group

FIG. 3-1 Functional chemical groups.

oxygen) as the Benedict's reagent is **reduced** (the loss of an oxygen). They will change from blue to green, orange, or red, indicating the varying amounts of the simple sugar.

Sucrose (table sugar) is an example of a disaccharide. It is made up of two monosaccharides, glucose and fructose, linked together by **dehydration synthesis.** Dehydration synthesis is the combining of two molecules by the removal of a water molecule. In reversal, a disaccharide may be broken down into two monosaccharides with the addition of water; this is known as **hydrolysis,** or condensation reaction.

Polysaccharides are long chains of covalently linked monosaccharides and have many different structures and functions. An example of a polysaccharide is **starch.** Its basic structure is due to **covalent bonding,** but the **conformation** (coiled structure) is due to **hydrogen bonding** and **ionic bonding.**

Starch can be distinguished from monosaccharides, disaccharides, and other polysaccharides by its reaction with the **iodine reagent,** which results in a distinctive blue-black color.

PROCEDURE A   MONOSACCHARIDE REAGENT TEST

1. Add 200 ml tap water to a 250 ml beaker. Use a hot plate to bring the water to a boil.
2. Label four clean test tubes 1 to 4. To labeled tubes add the following:

Tube 1   3 ml distilled water

Tube 2   3 ml glucose solution

Tube 3   3 ml sucrose solution

Tube 4   3 ml starch solution

3. Add 3 ml **Benedict's reagent** to each tube. Note color.
4. Place all four tubes into the boiling water bath for 3 minutes.
5. Using a test tube holder, remove test tubes to observe and record any color changes.

PROCEDURE B   DISACCHARIDE MODEL

1. Use the molecule model set to build the monosaccharide glucose.
2. Use the molecule model set to build the monosaccharide fructose.
3. Place the two molecules together so that a water molecule is released and a sucrose molecule is formed.

PROCEDURE C   STARCH AND IODINE REAGENT TEST

1. Label four clean test tubes 1 to 4. To labeled tubes add the following:

Tube 1   3 ml distilled water

Tube 2   3 ml glucose solution

Tube 3   3 ml sucrose solution

Tube 4   3 ml starch solution

2. Add 1 ml **iodine solution** to each tube and note color.
3. While holding the top of each test tube, mix the contents by gently swirling.
4. Record any color changes.

1. What are the three types of carbohydrates? Give examples of each.

2. Why are carbohydrates important to the human body?

3. Define dehydration synthesis.

4. To demonstrate dehydration synthesis draw the structural formula of the disaccharide maltose.

5. Define hydrolysis.

6. Demonstrate hydrolysis using the structural formula of the disaccharide sucrose.

7. Explain how the elements carbon, hydrogen, oxygen, nitrogen, sulfur, and phosphorus are important to the human body.

8. Explain how Benedict's reagent detects a monosaccharide.

9. Explain why an iodine reagent is specific to the polysaccharide starch.

10. What was the purpose of including test tube #1 containing distilled water in each of the experiments?

*Use all references and materials at your disposal to answer these review questions.

BIOLOGICAL MOLECULES
Lipids

4

OBJECTIVES

1 **Identify the presence of lipids in various substances.**

MATERIALS

filter paper disk
pencil
distilled water
flour solution
Sudan III reagent*

egg yolk solution
chicken soup
margarine
egg white
hair dryer

* **CAUTION: Harmful to skin**
and clothing. Rinse with
water if spillage occurs.

Lipids contain the atoms carbon, hydrogen, and oxygen but not in the 2 hydrogen : 1 carbon and 1 oxygen ratio that is seen in carbohydrates. Lipids are **nonpolar** molecules and may be classified according to their solubility. The most familiar lipid is fat, a **triglyceride.** This molecule consists of one **glycerol** molecule and three **fatty acid** molecules (Fig. 4-1).

The **carboxyl group** of one fatty acid molecule reacts with one of the **hydroxyl groups** of the glycerol molecule to form a triglyceride and water. (The resulting triglyceride is **nonpolar.**) With the carboxyl and hydroxyl groups bound up in the dehydration synthesis, the nonpolar molecule reacts with the **Sudan III reagent.** The color result is a bright orange.

Phospholipids are structurally similar to triglycerides; however, the third fatty acid molecule is replaced by a **polar phosphate group.** Therefore phospholipids are **hydrophobic** at one end and **hydrophilic** at the other end and behave as a detergent or soap. In the

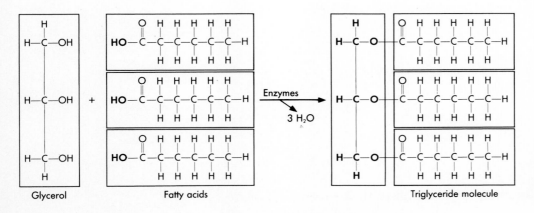

Glycerol   Fatty acids   Triglyceride molecule

FIG. 4-1 Production of one triglyceride from one glycerol molecule and three fatty acids.

phospholipid bilayer of the plasma membrane, the nonpolar ends of the phospholipid are "pushed" together due to the attraction of the two outside hydrophilic ends to water and other polar molecules.

PROCEDURE A   LIPID IDENTIFICATION

1. Using the letters w (water), f (flour), y (egg yolk), c (chicken soup), e (egg white), and m (margarine), label with a pencil on a filter paper disk around the edge so that the letters are spaced equally apart.
2. Add a small drop of each substance in the designated area.
3. Allow the paper to dry completely.
4. Place the filter paper disk in a Sudan III solution for 3 minutes.
5. Remove the filter paper disk with tweezers and rinse in tap water for 1 minute.
6. Use a hair dryer to dry completely.
7. Record the intensity of staining and record as follows:
   0   = no color
   +   = faint orange
   + + = definite orange

REVIEW QUESTIONS*

1. How are lipids different from carbohydrates?

2. How are lipids important to the human body?

3. Draw a glycerol molecule and three fatty acids. Circle the important functional groups.

4. On Fig. 4-1, draw a dotted line around the atoms released during dehydration synthesis.

5. Draw a phospholipid molecule and explain how it is different from a triglyceride.

6. How is the chemical structure of a phospholipid important to the cell membranes?

*Use all references and materials at your disposal to answer these review questions.

BIOLOGICAL MOLECULES
Proteins

OBJECTIVES

1 **Compare the movement of various amino acids on chromatography paper.**
2 **Detect the presence of protein in various substances.**

MATERIALS

chromatography paper
pencil
ruler
chromatography chamber
various amino acids
solvent (ethanol)
ninhydrin*
5 clean test tubes

test tube rack
water
white dog hair
chicken soup
egg white solution
fingernail clipppings
nitric acid*

* **CAUTION: Harmful to skin**
and clothing. Rinse with
water if spillage occurs.

KEY TERMS

Amino acid
Amino group
Carboxyl group
Catalytic protein
Contractile protein
Dehydration synthesis
Hydrolysis
Immunological protein
Ninhydrin
Peptide bond
Polypeptide
Protein
Regulatory protein
Structural protein
Transport protein
Variable group

The building blocks of proteins are the monomers called **amino acids.** They are joined together in **dehydration synthesis** by **peptide bonds** to form **peptides.** The term *peptide* is used when the molecule contains less than 100 amino acids, whereas the term *protein* designates large peptides made up of many hundreds of amino acids (Fig.5-1).

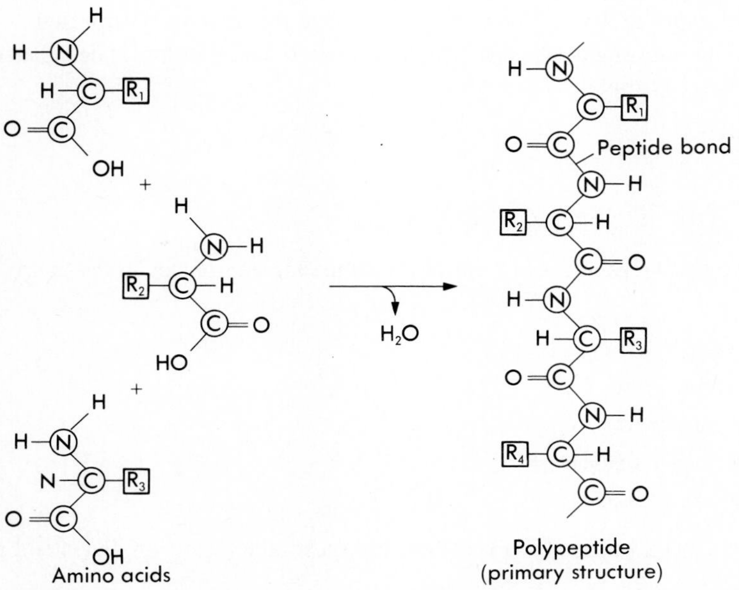

FIG. 5-1 Dehydration synthesis of a protein.

Amino acids

Polypeptide
(primary structure)

Proteins are responsible for various functions throughout the body: **structural proteins** make up a significant part of the cell and intercellar space; **regulator proteins,** such as hormones, function in various physiological processes; **contractile proteins** function in muscular movement; **immunological proteins** protect the body from microbial invasion; **transport proteins** move substances throughout the body; and **catalytic proteins,** such as enzymes, control biochemical reactions.

All amino acids contain a basic functional **amino group** ($-NH_2$) and **carboxyl group** ($-CO_2H$). Individual amino acids may be distinguished by a **variable group** (R group). The properties of proteins depend on the sequence of amino acids, the variable group, and the molecular size.

Proteins can be broken down into the component amino acid units by **hydrolysis,** usually in hot hydrochloric acid. The amino acids can then be separated by paper chromatography as a solvent flows over the sample.

Individual amino acids can be identified by their position on the chromatography paper. In most solvents, amino acids with alkyl chains and lighter amino acids migrate more rapidly than do those with polar groups on the side chains. To locate the amino acids on the paper chromatogram, use the reagent **ninhydrin.** A characteristic blue-violet color will appear.

Proteins may also undergo reactions characterized by amino acid side chains. For example, the yellowing of skin caused by nitric acid is a reaction of phenolic rings in the amino acid tyrosine present in skin proteins.

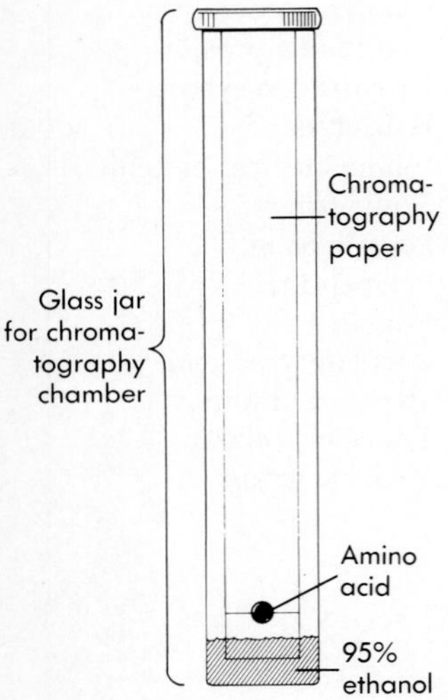

Glass jar for chromatography chamber

Chromatography paper

Amino acid

95% ethanol

FIG. 5-2 Chromatogram chamber.

PROCEDURE A   SEPARATION OF AMINO ACIDS USING A CHROMATOGRAM

1. Using a pencil, draw a line 1 cm from the bottom of a strip of chromatography paper. *Do not allow your finger to touch the paper.*
2. Below the 1 cm line designate which amino acids are to be used.
3. Using a capillary pipette, place a drop of the designated amino acid on the 1 cm line. Wait for the drop to dry; then add a second drop.
4. When the paper is completely dry, tape the top of the paper to the chromatography chamber. Add enough of a 95% ethanol solution to the chamber to cover the bottom of the paper, but not the amino acid. Carefully place the chromatography paper in the chamber (Fig. 5-2).
5. Wait 1 hour and remove the paper. (Because of possible time factors, this may need to be done by the instructor.) Allow it to dry overnight.
6. Saturate the paper carefully in ninhydrin.
7. Allow the paper to dry and measure the distance the amino acid traveled.
8. Compare the movement of your designated amino acid with other designated amino acids of class members.

PROCEDURE B   PROTEIN DETECTION

1. Label five test tubes and add 3 ml of the following substances in the appropriate test tube:

   Tube 1   water

   Tube 2   chicken soup

   Tube 3   white dog hair

   Tube 4   egg white solution

   Tube 5   fingernail clippings

2. Add 4 ml of concentrated nitric acid to each test tube. Wait 5 minutes and record the results.

1. What are the monomers of a protein molecule?

2. Explain the process by which peptide bonds are formed in a protein molecule.

3. List the types of proteins and give an example of each.

4. Draw an amino acid, then circle the carboxyl group, the amino group, and the variable group.

5. Why do amino acids separate on chromatography paper?

*Use all references and materials at your disposal to answer these review questions.

6 BIOLOGICAL MEMBRANES
Isotonic, hypotonic, and hypertonic environments

KEY TERMS

Glycolipid
Glycoprotein
Hydrophilic
Hydrophobic
Hypertonic
Hypotonic
Integrated protein
Intracellular membrane
Isotonic
Lipid tail
Osmosis
Phosphate group
Phospholipid bilayer
Plasma membrane
Selectively permeable
Surface proteins

OBJECTIVES

1 Use a model membrane to demonstrate the movement of substances across a membrane.
2 Using living cells, observe the effects of an isotonic, hypertonic, and hypotonic environment.
3 Sketch, label, and explain the function of the plasma membrane.

MATERIALS

500 ml beaker	plastic sandwich bag
raw egg	thread
3% acetic acid (or vinegar)	iodine solution
corn syrup	starch solution
balance scale	*Elodea* leaf
distilled water	10% salt solution
compound microscope	tissue paper
slide	coverslip

The function of the **plasma membrane** is to separate the cellular contents from the environment. This is achieved by a membrane that consists of a **phospholipid bilayer.** Each phospholipid layer contains a highly polar phosphate group and a very nonpolar lipid tail.

The **hydrophilic phosphate groups** interact with the water molecules inside and outside the cell; causing the **hydrophobic lipid tails** to aggregate. Since the lipid bilayer is impermeable to ions and highly polar molecules, specialized **integrated proteins** provide permeability for necessary ions and polar molecules. This gives the plasma membrane its **selectively permeable** characteristic. **Surface proteins** seem to provide a role in the maintenance of the human cell via cohesive and adhesive functions, transport functions, and receptor functions of the human cell. Embedded in the lipid core and extending to the surface of the membrane are **glycolipids** and **glycoproteins.** These molecules are important in cellular recognition.

Intracellular membranes compartmentalize specialized intracellular components and are permeable to the substances necessary for that specialization.

Osmosis is the movement of water through a selectively permeable membrane. This allows water, but not all solutes, to diffuse across. In an **isotonic environment** the surrounding solution contains the same ratio of solutes to water as does the cell. In a **hypotonic environment** the ratio of solute to water is less than that found in the cell, and water from the surrounding solution will move into the cell. In a **hypertonic environment** the solute to water ratio is greater than that in the cell, and water will move out from the cell into the surrounding solution (Fig. 6-1).

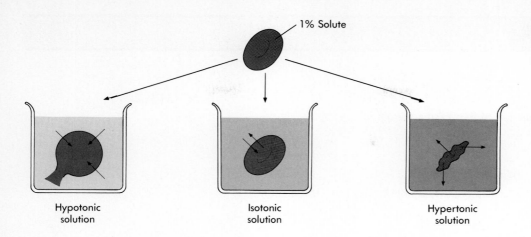

FIG. 6-1 Hypotonic, isotonic, and hypertonic solutions.

1% Solute

Hypotonic
solution

Isotonic
solution

Hypertonic
solution

PROCEDURE A   MODEL CELL MEMBRANE

1. To the plastic bag add 100 ml of a starch solution.
2. Fold one end over, squeeze out all the air, and tie it tightly with a piece of thread.
3. Weigh the bag and starch solution. Record the results.
4. Place the bag in a 250 ml beaker and add enough of the iodine solution to cover it.
5. On the following day, remove the bag and weigh the contents. Note any changes. Record the results.

PROCEDURE B   HYPOTONIC/HYPERTONIC ENVIRONMENTS

1. Place a raw egg (in the shell) in a beaker. Fill the beaker with 3% acetic acid. After 2 days, remove the egg and carefully rinse any remaining shell. Clean the beaker.
2. Weigh the egg and record the results.
3. Place the egg in the same beaker and add enough corn syrup to cover.
4. After 1 day remove, rinse, and weigh the egg. Record results.
5. After cleaning the beaker, place the egg in the beaker and cover the egg with distilled water.
6. After 1 day remove and weigh the egg. Record results.

PROCEDURE C   HYPOTONIC/HYPERTONIC ENVIRONMENTS

1. Make a wet mount of an *Elodea* leaf. Focus it on high power of a compound microscope. Draw observations.
2. While holding a small piece of tissue on one edge of the coverslip, add a drop or two of 10% NaCl solution to the opposite edge of the coverslip. The tissue will pull the solution through. Draw observations.
3. While watching in the microscope add a few drops of distilled water to the edge of the coverslip and pull it through with tissue on the opposite edge of the coverslip. This may need to be repeated several times.

1. Explain the structure and function of the plasma membrane.

2. Describe a selectively permeable membrane and how it may vary in function.

3. Define and give an example of the following terms:

   Diffusion

   Osmosis

   Filtration

4. What are the effects on red blood cells when placed in an isotonic solution? In a hypertonic solution? In a hypotonic solution?

5. Sketch and label a phospholipid bilayer.

6. How may certain polar molecules and ions permeate the plasma membrane bilipid layer?

*Use all references and materials at your disposal to answer these review questions.

BIOLOGICAL MEMBRANES
Active transport

OBJECTIVES	KEY TERMS

1 Use living cells to demonstrate active transport.
2 Use nonliving cells to demonstrate the need for energy during active transport.

Active transport
Neutral red

MATERIALS

two 1 g baker's yeast packets*
0.75% sodium carbonate
   solution
two 250 ml Erlenmeyer flasks
0.02% neutral red
hot plate

filter paper disk
small funnel
2 clean test tubes
test tube rack
3% acetic acid

* CAUTION: It is important
   that the yeast not be
   outdated.

Active transport is a carrier-mediated process by which substances, usually ions, are transported across a plasma membrane from a region of lower concentration to a region of higher concentration. This process is accomplished when the substance being transported attaches to a protein carrier molecule within or on the plasma membrane. The substance is then transported across the membrane and released to the opposite side. In some cases of active transport, one substance is exchanged for another. The energy necessary for this process is supplied by the breakdown of adenosine triphosphate (ATP) and adenosine diphosphate (ADP). In nonliving organisms no energy source would be available for active transport to occur.

In this investigation the substance being transported is a color indicator—**neutral red.** In an acidic environment the indicator is red, and in a basic environment the indicator is yellow. Since a living yeast cell is slightly acidic, for this investigation it will be placed in a basic solution of sodium carbonate to demonstrate active transport.

PROCEDURE A   ACTIVE TRANSPORT

1. Obtain two 1 g samples of baker's yeast.
2. In a 250 ml Erlenmeyer flask, place 1 g yeast and 25 ml of 0.75% sodium carbonate solution, and mix well. Boil this flask gently for 2 minutes and cool.
3. To this flask add 25 ml of 0.02% neutral red solution. Record the color of the suspension.
4. In a second flask place 25 ml of a 0.75% sodium carbonate solution and 1 ml of 0.02% neutral red. Record the color of the solution.

5. Add 1 g of yeast cells, swirl the flask, and watch for a color difference compared with the first flask. Record the final color of each flask.
6. Filter a portion of the contents of the first flask into a test tube. Record the color of the yeast cells on the filter paper and the filtrate solution in the test tube.
7. Filter a portion of the contents of the second flask into a test tube. Record the color of the cells on the filter paper and the filtrate solution in the test tube.
8. To prove that neutral red was actively transported into the living yeast cells and not the dead yeast cells, add 5 ml of 3% acetic acid to the filtrate solution. Record the results.

REVIEW QUESTIONS*

1. Compare passive transport and active transport.

2. Define exocytosis and endocytosis.

3. What is the difference between phagocytosis and pinocytosis?

4. How are carrier molecules important in active transport?

5. What did boiling the yeast do in this experiment and why?

*Use all references and materials at your disposal to answer these review questions.

CELL STRUCTURE MICROGRAPHS

8

OBJECTIVES

1 Identify cellular structures using transmission electron micrographs.
2 Explain the function of cellular structures.
3 Compare normal cell structure micrographs with abnormal cell structure micrographs.

KEY TERMS

Autophagy
Centrioles
Cilia
Cisternae
Cristae
Cytoskeleton
Endoplasmic reticulum
Flagella
Golgi complex
Lysosomes
Microtubrical lattice
Mitochondria
Nuclear pores
Nucleoli
Nucleus
Organelle
Phospholipid bilayer
Plasma membrane
Ribosomes
Rough endoplasmic reticulum
Selectively permeable
Smooth endoplasmic reticulum
Transmission electron microscope

MATERIALS

transmission electron micrographs (normal and abnormal cell structures)

Advancements in the **transmission electron microscope** have made it possible for scientists to more thoroughly study the function of cellular components. Identification of normal cellular structure is important and is used when comparing possible abnormal structures.

Animal cell components are classified into three basic groups: the plasma membrane, the cytoskeleton, and the organelles.

The **plasma membrane** is a boundary between the intracellular components and the extracellular materials. It is a single unit membrane composed of a **phospholipid bilayer** and proteins (Fig. 8-1). Since the cell must interact with the environment, certain substances are allowed to enter or exit the cell and other substances are restricted. Therefore the plasma membrane is considered to be **selectively permeable.**

FIG. 8-1 Electron micrograph of plasma membrane *(arrows).*

23

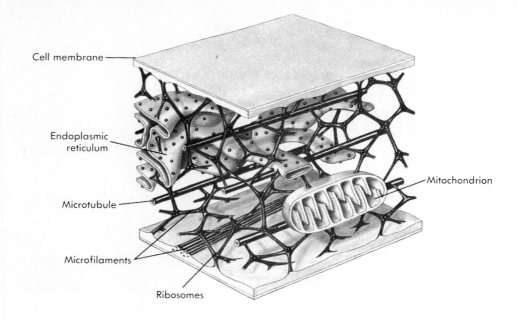

Cell membrane

Endoplasmic reticulum

Mitochondrion

Microtubule

Microfilaments

Ribosomes

FIG. 8-3 Nuclear envelope.
A, Electron micrograph.
B, Typical structure.

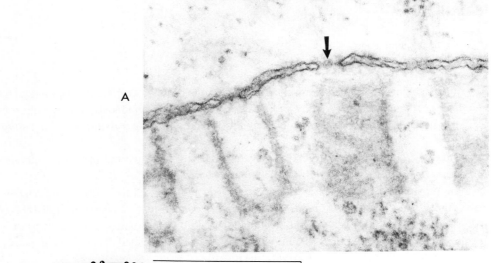

A

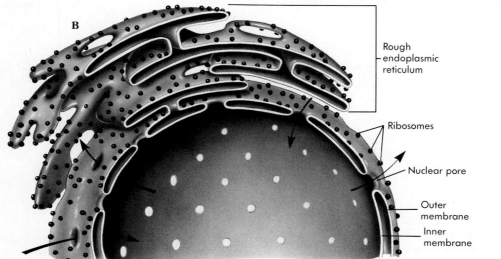

B

Rough endoplasmic reticulum

Ribosomes

Nuclear pore

Outer membrane

Inner membrane

HISTOLOGY

PLATE 1   Electron micrograph of interphase.

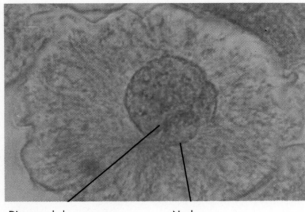

Dispersed chromosomes   Nuclear
form chromatin   envelope

PLATE 2   Electron micrograph of mitosis. **A,** prophase, **B,** metaphase, **C,** anaphase, **D,** telophase.

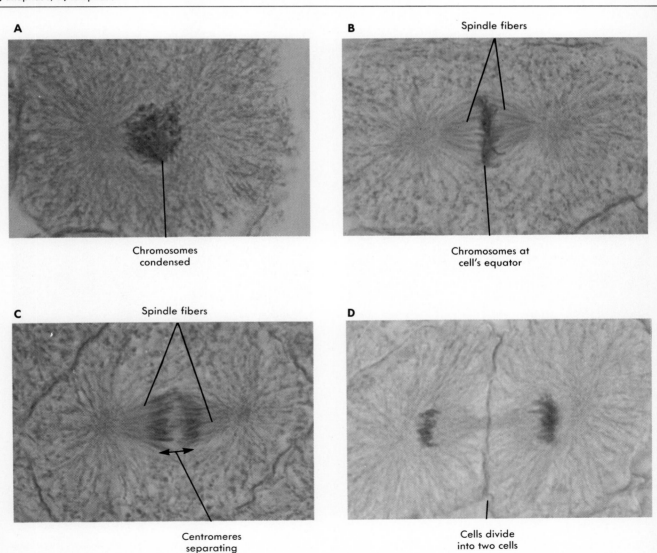

A

Chromosomes
condensed

B

Spindle fibers

Chromosomes at
cell's equator

C

Spindle fibers

Centromeres
separating

D

Cells divide
into two cells

PLATE 3 **A,** Simple squamous epithelium (surface view). **B,** Simple cuboidal epithelium (human). **C,** Simple columnar epithelium (human uterus).

A

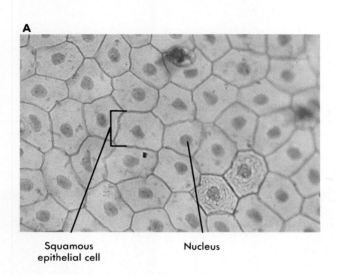

Squamous
epithelial cell

Nucleus

B

Nucleus

Basement membrane

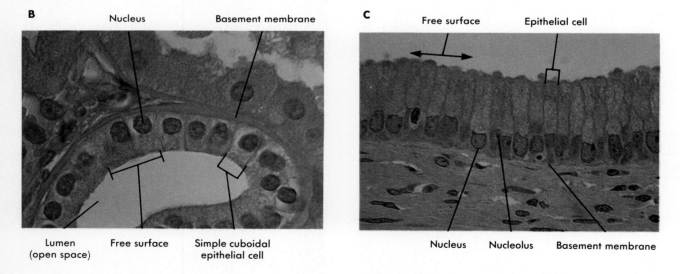

Lumen
(open space)

Free surface

Simple cuboidal
epithelial cell

C

Free surface

Epithelial cell

Nucleus

Nucleolus

Basement membrane

PLATE 4 **A,** Stratified squamous epithelium (human vagina). **B,** Stratified cuboidal epithelium (ovarian follicle). **C,** Stratified columnar epithelium (human). **D,** Pseudostratified epithelium.

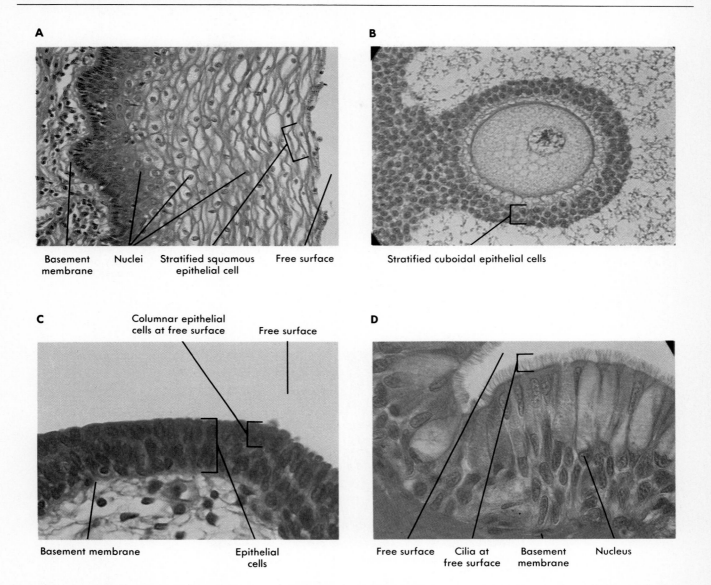

A

Basement membrane Nuclei Stratified squamous epithelial cell Free surface

B

Stratified cuboidal epithelial cells

C

Columnar epithelial cells at free surface Free surface

Basement membrane Epithelial cells

D

Free surface Cilia at free surface Basement membrane Nucleus

PLATE 5 Goblet cell.

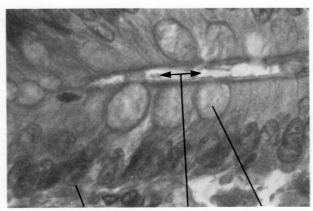

Basement membrane Free surface Goblet cell

PLATE 6 **A,** Areolar (loose) connective tissue. **B,** Dense fibrous (irregular collaginous) connective tissue. **C,** Dense regular (elastic) connective tissue. **D,** Adipose tissue. **E,** Reticular connective tissue.

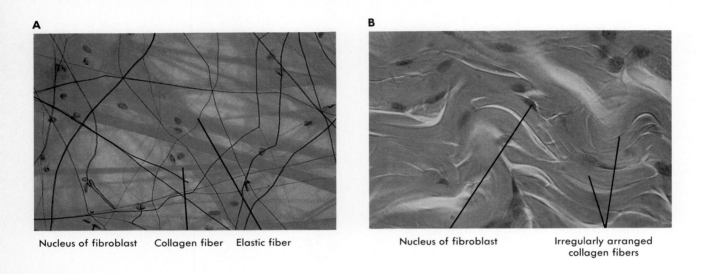

A

B

Nucleus of fibroblast   Collagen fiber   Elastic fiber

Nucleus of fibroblast   Irregularly arranged
collagen fibers

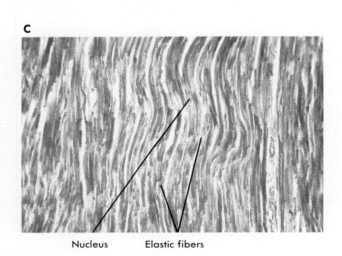

C

Nucleus   Elastic fibers

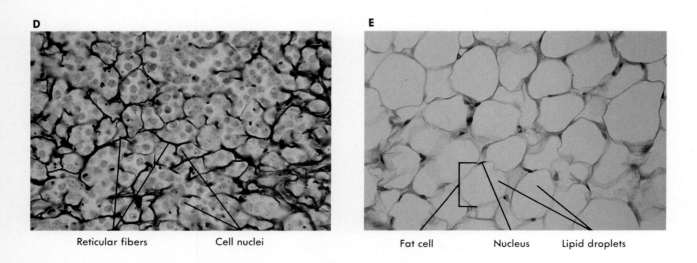

D

E

Reticular fibers   Cell nuclei

Fat cell   Nucleus   Lipid droplets

PLATE 7   **A,** Hyaline cartilage. **B,** Fibrocartilage. **C.** Elastic cartilage.

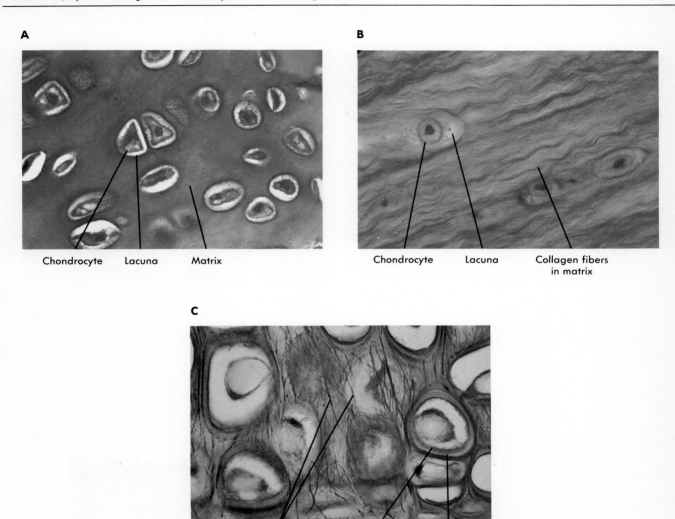

A

Chondrocyte   Lacuna   Matrix

B

Chondrocyte   Lacuna   Collagen fibers
in matrix

C

Elastic fibers   Chondrocyte   Lacuna
in matrix

PLATE 8   **A,** Cancellous bone. **B,** Compact bone.

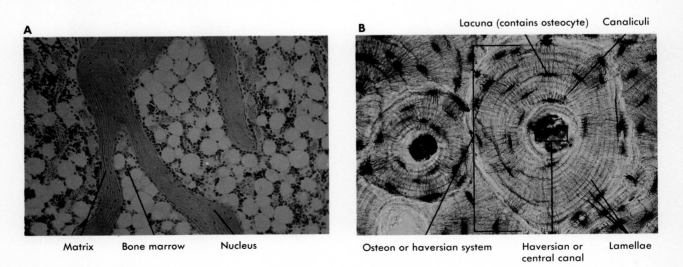

A

Matrix   Bone marrow   Nucleus

B

Lacuna (contains osteocyte)   Canaliculi

Osteon or haversian system   Haversian or   Lamellae
central canal

PLATE 9 Skin histology.

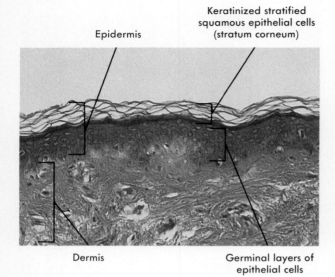

Epidermis

Keratinized stratified
squamous epithelial cells
(stratum corneum)

Dermis

Germinal layers of
epithelial cells

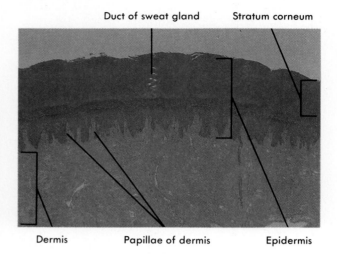

Duct of sweat gland

Stratum corneum

Dermis

Papillae of dermis

Epidermis

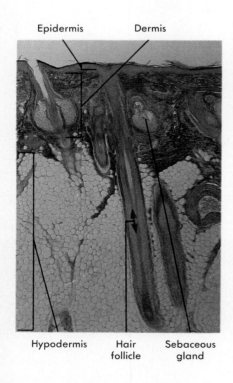

Epidermis

Dermis

Hypodermis

Hair
follicle

Sebaceous
gland

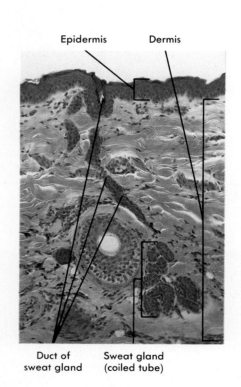

Epidermis

Dermis

Duct of
sweat gland

Sweat gland
(coiled tube)

PLATE 10   Skeletal muscle histology.

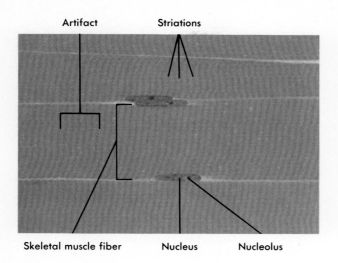

Artifact   Striations

Skeletal muscle fiber   Nucleus   Nucleolus

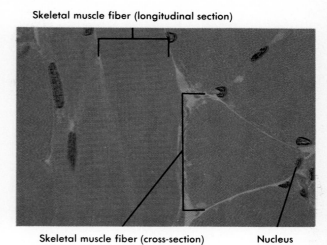

Skeletal muscle fiber (longitudinal section)

Skeletal muscle fiber (cross-section)   Nucleus

PLATE 11   Histology. **A,** Unipolar neuron. **B,** Multipolar neuron. **C,** Neuroglia.

A

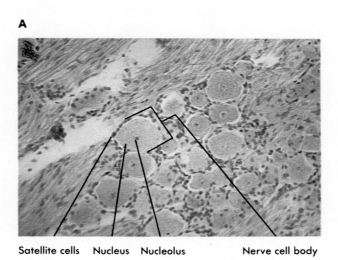

Satellite cells   Nucleus   Nucleolus   Nerve cell body

B

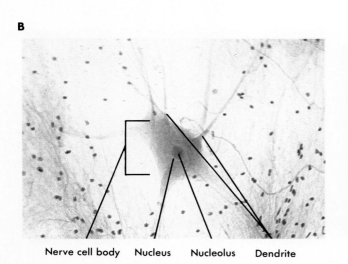

Nerve cell body   Nucleus   Nucleolus   Dendrite

C

Neuroglial cells (neuroglia)

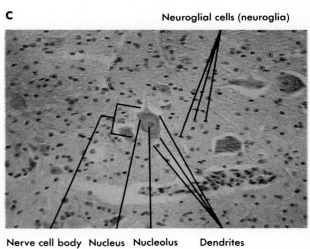

Nerve cell body   Nucleus   Nucleolus   Dendrites

A

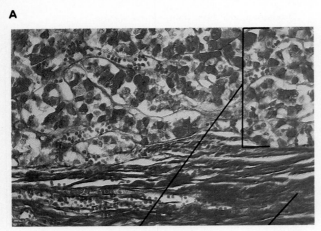

Pars distalis   Capsule

B

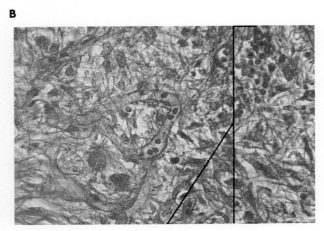

Pars nervosa

C

Parafollicular cells   Follicular cells

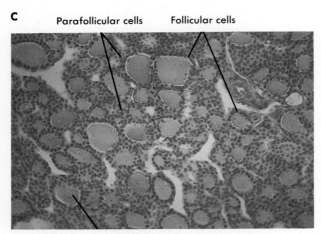

Follicle (containing thyroglobulin)

D

Adrenal medulla

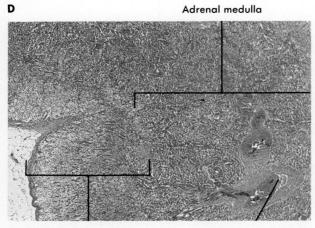

Adrenal cortex   Medullary veins

E

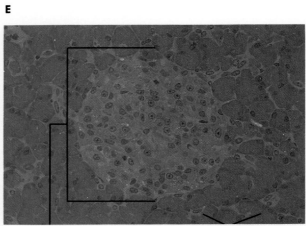

Endocrine portion of pancreas   Exocrine portion of pancreas
(islet of Langerhans)

PLATE 13   Formed elements (blood cells).

RED BLOOD CELLS   PLATELETS

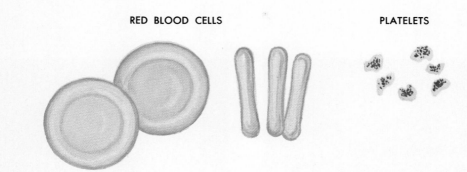

WHITE BLOOD CELLS (LEUKOCYTES)

GRANULAR LEUKOCYTES

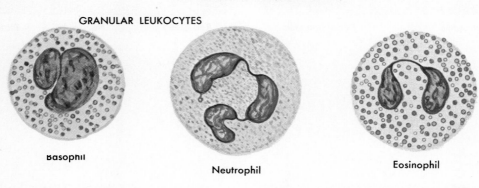

Basophil

Neutrophil

Eosinophil

NONGRANULAR
LEUKOCYTES

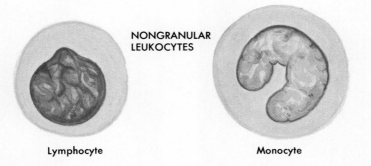

Lymphocyte

Monocyte

PLATE 14 **A** and **B**, Blood smear. **C**, Lymphocyte. **D**, Neutrophil.

A

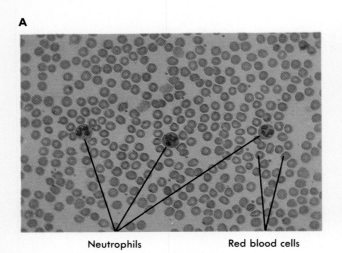

Neutrophils   Red blood cells

B

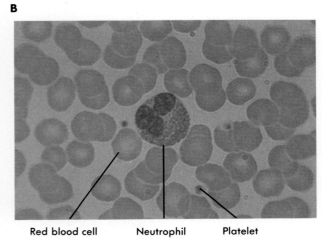

Red blood cell   Neutrophil   Platelet

C

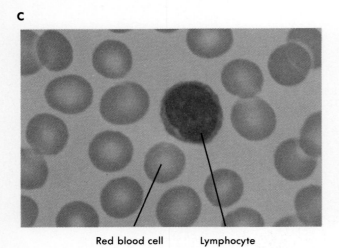

Red blood cell   Lymphocyte

D

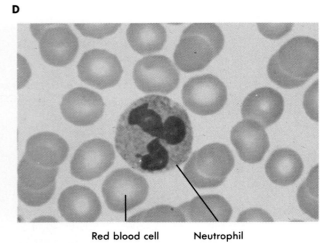

Red blood cell   Neutrophil

PLATE 15 Cardiac muscle tissue.

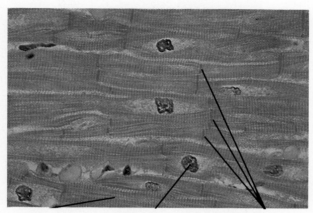

Note striations   Nucleus of cardiac   Intercalated discs
                       muscle cell   (special junctions
                                      between cells)

PLATE 16 Smooth muscle tissue.

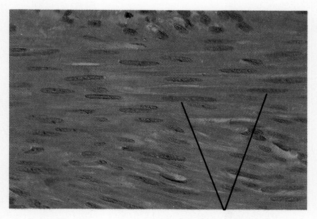

Nuclei of smooth muscle

PLATE 17 **A,** Lymph node histology. **B,** Spleen histology. **C,** Thymus histology.

A Lymphocytes in lymph node   Capsule

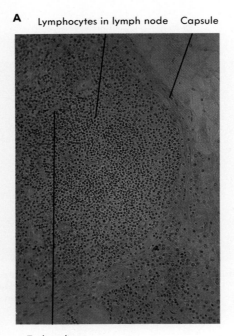

Trabeculae

B Red pulp   Blood vessels   White pulp

C   Vessel   Cell nuclei   Medulla

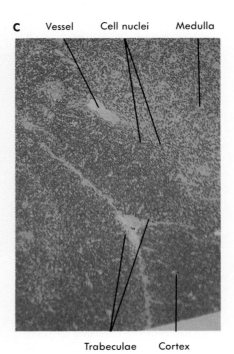

Trabeculae   Cortex

PLATE 18 Seminiferous tubule.

Basement membrane

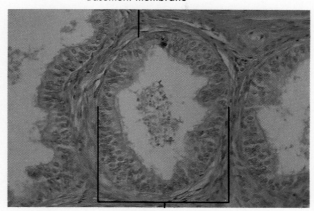

Seminiferous tubule

PLATE 19 Graafian follicle (cat).

Cumulus mass Theca externa Graafian or
Theca interna mature ovarian follicle

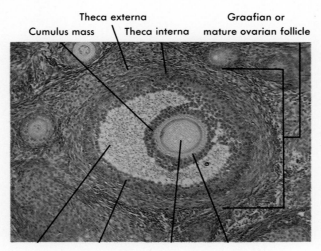

Antrum Granulosa cells Oocyte Zona pellucida

The **cytoskeleton** is a **microtubrical lattice** inside the plasma membrane and outside the nucleus (Fig. 8-2). It is responsible for cell shape and intracellular support and transport.

Organelles are membrane-bound parts of the cell and specialize in providing the synthesis, production, storage, and transport of chemicals necessary for growth, maintenance, repair, and control of the cell. Following are some examples of cell organelles:

Nucleus

The nucleus is a large organelle bound by a double membrane, which is pinched off in areas to form **nuclear pores** (Fig. 8-3). This organelle contains the genetic information as well as an aggregation of RNA called the **nucleolus.**

Endoplasmic reticulum

The endoplasmic reticulum is an extension of the nuclear double membrane. The lumina, or spaces, formed between these membranes are known as **cisternae.** The membranes fold back and forth upon one another. **Rough endoplasmic reticulum** has associated **ribosomes** that are necessary for protein synthesis. The ribosomes look like small dots along the endoplasmic reticulum (Fig. 8-4). **Smooth endoplasmic reticulum** has no associated ribosomes and is primarily involved in the synthesis of carbohydrates, lipids, and other nonprotein products.

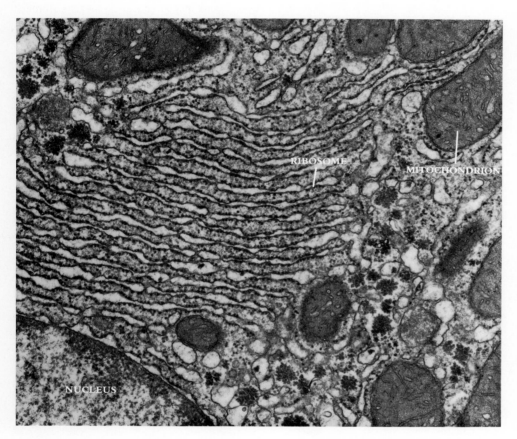

A

FIG. 8-4 Rough endoplasmic reticulum. **A,** Electron micrograph. **B,** Typical structure.

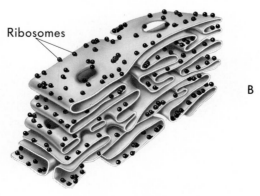

Ribosomes

B

Golgi complex

A series of membranous sacs stacked on one another form the Golgi complex. Substances synthesized along endoplasmic reticulum are encased by a membrane in the Golgi complex. These vessicles pinch off and either remain in the cell or are carried to the plasma membrane. The membranes of the vessicles fuse with the plasma membrane and the content is secreted (Fig. 8-5).

Lysosomes

Lysosomes are small rounded membrane-bound sacs. They contain hydrolytic enzymes used for digesting various aging components within the cell. This organelle is known as the "suicide bag," since it will destroy an aging or nonfunctioning cell by **autophagy** ("self-eating") (Fig. 8-6).

Mitochondria

Structurally, mitochondria can be tubular, elliptical, or circular with a double membrane. The inner membrane has many folds, called the **cristae.** Mitochondria convert the chemical-bond energy of nutrients and store it in the high-energy bonds of ATP (Fig. 8-7).

Centrioles

Centrioles are paired cylinders at right angles to one another. Through the electron microscope a cross section of a centriole is seen as nine sets of three microtubules. The centrioles seem to be the microtubular organization centers of the cell. Their role during cellular division is still unclear (Fig. 8-8).

Cilia and Flagella

Cilia and flagella are structures that provide cellular movement. Although cilia are numerous short cellular projections and flagella are few, long cellular projections, they are structurally the same. Through the electron microscope, both cilia and flagella cross section arrangements are seen as nine pairs of microtubules with two single microtubules in the center. A longitudinal section is seen as three long strands (Fig. 8-9).

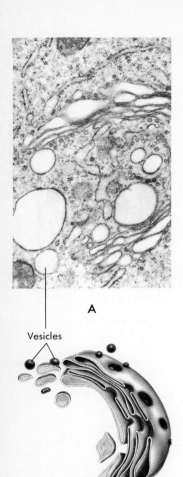

A

Vesicles

B

FIG. 8-5 Golgi complex.
A, Electron micrograph.
B, Typical structure.

FIG. 8-6 Electron micrograph of lysosomes *(L).*

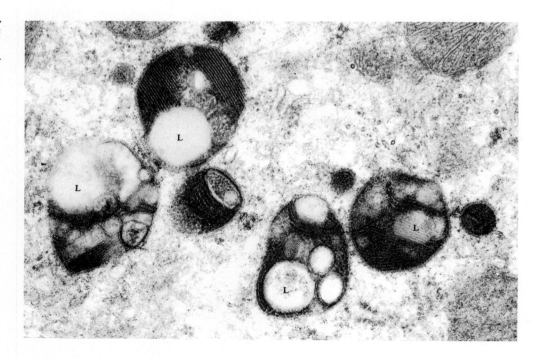

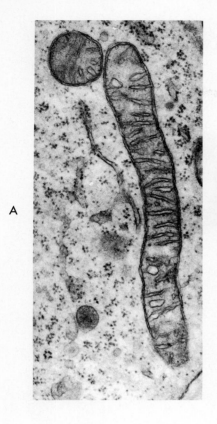

A

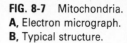

FIG. 8-7 Mitochondria.
A, Electron micrograph.
B, Typical structure.

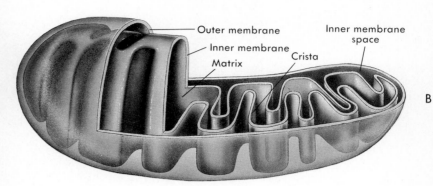

Outer membrane   Inner membrane space

Inner membrane   Crista

Matrix

B

FIG. 8-8 Centrioles. A, Electron micrograph. B, Typical structure.

Microtubule triplet

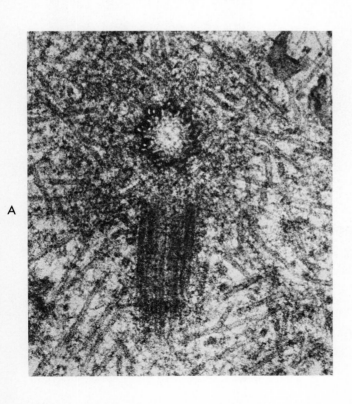

A

B

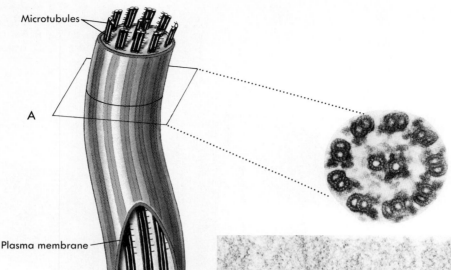

Microtubules

A

Plasma membrane

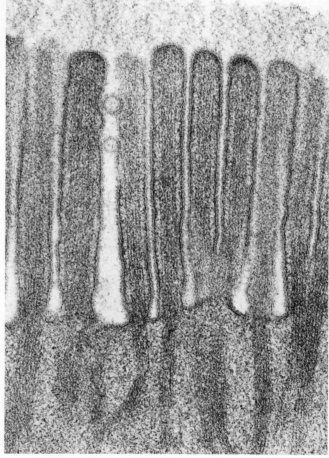

B

FIG. 8-9 Cilia and flagellae.
A, Cross section.
B, Longitudinal section.

PROCEDURE A ELECTRON MICROGRAPH CELL STRUCTURE

1. Use the transmission electron micrographs to identify the various cellular structures and explain the function of each.
2. Using the micrograph, draw a picture of the various structures. Label each structure and define its function.

PROCEDURE B NORMAL AND ABNORMAL CELL STRUCTURE COMPARISON

1. Use micrographs of cancerous cells to compare the possible differences in various structures.

1. Compare the compound microscope and the transmission electron microscope.

2. Explain the function of the plasma membrane and how this function is achieved.

3. Describe the structure of the cytoskeleton and explain how it is important to the cellular function.

4. Describe the pathway of a secretory enzyme once it has been synthesized at the rough endoplasmic reticulum.

5. How are lysosomes important to aging or nonfunctioning cells?

6. Compare the ultrastructure of centrioles and cilia.

7. How does the nuclear membrane differ from the plasma membrane? Why is this important?

*Use all references and materials at your disposal to answer these review questions.

CELLULAR METABOLISM

OBJECTIVES

1 **Measure the amount of carbon dioxide production as an aerobic organism undergoes the metabolic process.**

MATERIALS

four 250 ml beakers
distilled water
phenolphthalein solution

graduated cylinder (50 ml)
0.0025 M sodium hydroxide
  (NaOH)

* **CAUTION: Harmful to skin**
  and clothing. Rinse with
  water if spillage occurs.

When the chemical energy bonds of carbohydrate, fat, and protein molecules are broken down, the energy released is stored in the bonds of ATP molecules. The stored energy may then be used for various ongoing cellular processes. A familiar cellular fuel molecule, glucose, is one source of chemical energy that can be transferred to high-energy bonds of ATP.

Glucose **metabolism** is a series of oxidation-reduction reactions. **Oxidation** is the gain of oxygen, loss of hydrogen, or loss of an electron by an atom, ion, or molecule. **Reduction** is the loss of oxygen, gain of hydrogen, or gain of an electron by an atom, ion, or molecule. This process may be divided into the following three sequential parts and one intermediate step:

Glycolysis occurs in the cytoplasm, where a six-carbon glucose molecule is broken down to two three-carbon **pyruvate** molecules. The energy released is stored in the phosphate bonds of ATP and in the reduction (the addition of a hydrogen atom or electron) of the carrier molecule, NAD. NAD is reduced to $NADH^+H^+$.

In the **intermediate step,** from each three-carbon pyruvate molecule one carbon is released in the form of carbon dioxide. The released energy from those bonds reduces NAD to $NADH^+H^+$.

Krebs cycle occurs in the matrix of the mitochondria, where each two-carbon molecule joins a four-carbon molecule. During the rearrangement of this six-carbon molecule, carbon and oxygen atoms in the form of carbon dioxide are released. The released energy from those bonds reduces NAD to $NADH^+H^+$ and FAD to $FADH_2$. Some of the released energy is also directly stored in ATP molecules.

The **electron transport chain** occurs in the cristae and matrix of the mitochondria. A series of oxidation/reduction reactions transfer energy from the carrier molecules to ATP in a process called **oxidative phosphorylation.**

The production and release of carbon dioxide by an organism in water will form carbonic acid. Phenolphthalein is a color indicator. At pH 8-10 the indicator is pink.

PROCEDURE A

1. Label four beakers and fill each beaker with 100 ml DH_2O and one of the following organisms:

   Beaker 1   distilled water only

   Beaker 2   distilled water and a large goldfish

   Beaker 3   distilled water and two snails

   Beaker 4   distilled water and 5 cm *Elodea*

2. Cover each beaker with aluminum foil.
3. Allow the organisms to remain in the beaker for 30 minutes.
4. Remove the organisms and place each in a 50 ml graduated cylinder containing 40 ml tap water. Record the difference (the volume of the organism). Return the organisms to the proper containers.
5. To each beaker add 25 ml phenolphthalein solution.
6. While stirring, add NaOH drop by drop to each solution until there is a definite color change. Record the number of drops for each solution.
7. Calculate CO_2 production:

$$CO_2 \text{ production} = \frac{(\text{ml NaOH of experimental} - \text{ml NaOH of control}) \times (0.0025 \text{ MNaOH})}{\text{volume of organism (ml)} \times \text{time (hr)}}$$

1. Define metabolism, catabolism, and anabolism.

2. Where does the energy come from or from where is the energy obtained, that is necessary to catabolize a glucose molecule?

3. Where does glycolysis occur?

4. Compare anaerobic and aerobic respiration.

5. What is the end product of glycolysis?

7. What part of metabolism occurs in the mitochondria?

8. What is the purpose of the Krebs cycle in cellular metabolism?

9. What happens to the transferred energy from the Krebs cycle?

10. What is the purpose of the electron transport chain?

11. Why is oxygen essential for most living organisms?

*Use all references and materials at your disposal to answer these review questions.

DNA MOLECULES AND DNA REPLICATION

10

OBJECTIVES

1 Describe the structure of a nucleotide.
2 Demonstrate the covalent and hydrogen bonding in a DNA molecule.
3 Demonstrate the synthesis of a DNA molecule.

KEY TERMS

Adenine
Catalytic protein
Complementary pair
Cytosine
Deoxyribose
DNA (deoxyribonucleic acid)
Guanine
Histones
Nucleotide
Phosphate
Ribose
Thymine

MATERIALS

96 large paper clips (represent deoxyribose sugar)
96 colored paper clips (represent phosphate)
24 blue pipe cleaners (represent adenine)
24 red pipe cleaners (represent thymine)

24 green pipe cleaners (represent cytosine)
24 yellow pipe cleaners (represent guanine)
4 pieces of 1 in x 3 in Styrofoam

A **chromosome** consists of a double stranded helical **DNA (deoxyribose nucleic acid) molecule** and associated proteins called **histones.** The DNA polymer is made up of monomers called **nucleotides.** A DNA nucleotide consists of the following:

1. A cyclic five-carbon sugar molecule called a **deoxyribose.**
2. One of four nitrogenous bases, **adenine** (A), **thymine** (T), **cytosine** (C), and **guanine** (G). They are attached to the $1'$ carbon of the sugar.
3. A **phosphate** is attached to the $5'$ carbon of the sugar.

Nucleotides on each of the two strands of DNA are linked together by covalent bonds between the sugar and phosphate. The two strands link together as hydrogen bonds form between **complimentary bases** (adenine $=$ thymine, cytosine $=$ guanine).

Before a cell divides, the DNA replicates so that each new cell receives the same information as the parental cell.

The following represents a hypothetical double-stranded DNA molecule. Each letter represents the nitrogenous base of a nucleotide.

T-A-C-A-A-G-C-T-T-C-T-G-G-C-T-A-C-A-G-G-C-A-T-C
A-T-G-T-T-C-G-A-A-G-A-C-C-G-A-T-G-T-C-C-G-T-A-G

PROCEDURE A — SYNTHESIS OF A DNA MOLECULE

1. Link one large paper clip to a colored paper clip, then twist on a blue pipe cleaner to the large paper clip. This represents the first nucleotide of the DNA molecule.
2. Link nucleotides of the top base sequence together until the strand is complete. Begin linking nucleotides of the bottom sequence. Twist the pipe cleaners of complementary bases (representing hydrogen bonds) together as each new nucleotide is added.
3. Place the two ends of the double-stranded DNA molecule into two pieces of Styrofoam. Hold one end while your lab partner holds the other end; twist the strands of DNA. A double-stranded helical DNA model is now complete.

PROCEDURE B — REPLICATION OF A DNA MOLECULE

1. Beginning at the first complementary pair of the DNA molecule, untwist the hydrogen bond.
2. After untwisting each hydrogen bond, begin adding a new complimentary nucleotide to each of the original DNA strands. You will end up with two double-stranded DNA molecules, each a replica of the original double strand.

REVIEW QUESTIONS*

1. What is a nucleic acid?

2. What is the monomer of a nucleic acid?

3. Draw and label a DNA monomer.

4. Draw and define the terms *purines* and *pyrimidines*.

5. Explain how a DNA molecule is assembled.

6. How are complementary bases bonded? What is the importance of this type of bonding?

*Use all references and materials at your disposal to answer these review questions.

CELL CYCLE

11

OBJECTIVES

1 Identify and explain each phase of the cell cycle.
2 Demonstrate interphase, mitosis, and cytokinesis of the cell cycle.

MATERIALS

animal mitosis slide
large sheet of paper
   (60 cm × 60 cm;
   24 in x 24 in)

2 different colors of clay
pencil

KEY TERMS

Anaphase
Asters
Cell cycle
Centrioles
Centromere
Chromatin
Chromosome
Cytokinesis
G-1 phase
G-2 phase
Histone
Interphase
Kinetochore
Metaphase
Metaphase plate
Mitosis
Mitotic spindle
Nonhistone
Nuclear membrane
Nucleoli
Prophase
S-phase
Telophase

The growth, chromosomal replication, nuclear division, and cytoskeletal division of a cell are called the **cell cycle.** It is a continuous process classified by three stages: **interphase, mitosis,** and **cytokinesis.**

During the cell cycle, the terminology for the physical characteristics of the genetic information varies according to what stage it's in. The term **chromatin** is used to describe the loosely coiled DNA molecule with its associated proteins, called **histones,** and nonproteins, called **nonhistones.** Tightly coiled or condensed chromatin is referred to as **chromosomes.**

The first stage of the cell cycle is interphase (Color Plate 1). It contains three subphases crucial to the preparation of the cell for mitosis. During the **G-1 (gap one) phase,** the DNA is loosely coiled during the transcribing and translating of the proteins necessary for growth. During the **S (synthesis) phase,** DNA and chromosomal proteins are replicating. At this point the replicated chromatin begins to condense, forming a chromosome, and each individual strand is referred to as a **chromatid.** During the **G-2 (gap two) phase,** synthesis and organization of microtubules into the **mitotic spindle** (spindle fibers) occurs.

The second stage of the cell cycle, when nuclear information is divided, is known as **mitosis.** Mitosis is subdivided into **prophase, metaphase, anaphase,** and **telophase.** During prophase the **centrioles** migrate to opposite poles. The visible microtubular bodies around the centrioles are called **asters.** Their function is not known.

At this stage chromosomes are tightly coiled. An area where replication has not been fully completed is the **centromere.** Here the mitotic spindles attach at a site called the **kinetochore.** Also during prophase the **nucleoli** disperse and the **nuclear membrane** becomes disorganized and is no longer evident (Color Plate 2, *A*).

In the second phase of mitosis, metaphase, the sister chromatids line up along the metaphase plate. This ensures equal numbers and kinds of chromosomes in each daughter cell (Color Plate 2, *B*).

During anaphase, the chromatids separate (the centromere divides) and move toward the opposite poles (Color Plate 2, *C*).

In the last phase of mitosis, telophase, nuclear membrane reorganization occurs, the nucleoli reappear, and the chromosomes uncoil (Color Plate 2, *D*).

The cell cycle concludes with cytokinesis, the process in which the plasma membrane divides the cytoskeletal information. At the end of cytokinesis, there are two cells identical to the original cell. These cells continue the cell cycle and enter into interphase.

PROCEDURE A MICROSCOPIC OBSERVATION OF ANIMAL MITOSIS

1. Obtain an animal mitosis slide and focus it on high power under the compound microscope.
2. Locate each phase of the cell cycle beginning with interphase.
3. Explain what is occurring during each phase.

PROCEDURE B MODEL OF ANIMAL MITOSIS

1. Using a large sheet of paper draw and label a circle to represent the plasma membrane of each phase and subphase of the cell cycle.
2. Using a pencil, draw in the changes of the mitotic spindles, nuclear membrane, and centrioles.
3. Using 2 different colors, obtain enough clay to form chromatin strands and their replicated copy.
4. Beginning with interphase, roll out the original chromatin, then make the replicated copy. Pinch the clay together at the centromere.
5. Moving the same clay throughout each phase of the cell cycle drawn on the paper, demonstrate cellular activity at each phase. Coil and uncoil the clay at the appropriate phases.

1. Define the following terms:

   Chromatin

   Chromosome

   Chromatid

   Centromere

   Kinetochore

   Asters

2. Explain the importance of the synthesis phase of interphase.

3. Explain the importance of metaphase.

4. What is the function of the mitotic spindles?

5. Draw, label, and describe all the events that occur in each stage of the cell cycle.

*Use all references and materials at your disposal to answer these review questions.

12

DNA TRANSCRIPTION AND TRANSLATION

KEY TERMS

Adenine
Anticodon
Codon
Cytosine
Guanine
Messenger RNA (mRNA)
Nontranscribed strand
Peptide bond
Phosphate
Ribose
Ribosomal RNA (rRNA)
Ribosome
Transcribing strand
Transcription
Transfer RNA (tRNA)
Translation
Uracil

OBJECTIVES

1 Demonstrate transcription of genetic information by the synthesis of a messenger RNA molecule.

2 Demonstrate translation of genetic information.

MATERIALS

24 large paper clips (represent sugar)
24 colored paper clips (represent phosphate)
6 blue pipe cleaners (represent adenine)
6 orange pipe cleaners (represent uracil)
6 green pipe cleaners (represent cytosine)
6 yellow pipe cleaners (represent guanine)

7 small boxes (represent tRNA) with appropriate anticodons on each box
7 small styrofoam balls (represent amino acids) with each labeled amino acid corresponding to the codons
6 tooth picks (represent peptide bonds)

The process by which a gene makes a protein involves the **transcription** (a copy) and **translation** (change from one state to another) of the DNA code. Only one strand of DNA called the **transcribing strand** is used. The complementary DNA strand is called the **nontranscribed strand.**

Transcriptions include the synthesis of single-stranded complementary RNA nucleotides to form a **messenger RNA (mRNA)** molecule. The polymer RNA (ribose nucleic acid) is made up of nucleotide monomers. An RNA nucleotide is similar to a DNA nucleotide; however it consists of the following:

1. A cyclic five-carbon sugar molecule called a **ribose.**
2. One of four nitrogenous bases: **adenine** (A); **uracil** (U), present instead of thymine in DNA; **cytosine** (C); and **guanine** (G). They are attached to the 1′ carbon of the sugar.
3. A **phosphate** is attached to the 5′ carbon of the sugar.

A series of three nucleotide bases called **codons** will code for a specific amino acid.

Once an mRNA molecule is transcribed in the nucleus, it moves out to the cytoplasm along with the rough endoplasmic reticulum. Also in the cytoplasm are **transfer RNA (tRNA)** molecules. Their unique structure contains an **anticodon** of three complementary nucleotide bases at one loop and a site that will transfer a specific amino acid to the mRNA codon.

Translation occurs with the aid of **ribosomes.** They are two subunits composed of

ribosomal RNA (rRNA) and proteins. Ribosomes temporarily hold the mRNA and tRNA, while amino acids are linked together by **peptide bonds** to form the coded protein.

There are 64 codon combinations; therefore more than one codon codes for an amino acid. Translation always begins with the codon AUG. There are three codons that do not code for amino acids but represent a stop signal for transcription and translation to end.

PROCEDURE A   DNA TRANSCRIPTION AND TRANSLATION

1. Use the top DNA strand from the box on p. 45 to transcribe an mRNA molecule (nucleotide by nucleotide) with paper clips and pipe cleaners.
2. Using the mRNA codon sequence (Table 12-1), determine which amino acid to bring in with the appropriate tRNA.
3. Place the first amino acid in the corresponding tRNA box. Use the anticodon site to match up with the codon.
4. As the first and second amino acids are brought in, use a tooth pick to link the amino acids.
5. The last codon will represent a stop signal.

TABLE 12-1   THE GENETIC CODE

FIRST LETTER	SECOND LETTER				THIRD LETTER
	U	C	A	G	
U	Phenylalanine	Serine	Tyrosine	Cysteine	U
	Phenylalanine	Serine	Tyrosine	Cysteine	C
	Leucine	Serine	Stop	Stop	A
	Leucine	Serine	Stop	Tryptophan	G
C	Leucine	Proline	Histidine	Arginine	U
	Leucine	Proline	Histidine	Arginine	C
	Leucine	Proline	Glutamine	Arginine	A
	Leucine	Proline	Glutamine	Arginine	G
A	Isoleucine	Threonine	Asparagine	Serine	U
	Isoleucine	Threonine	Asparagine	Serine	C
	Isoleucine	Threonine	Lysine	Arginine	A
	(Start) Methionine	Threonine	Lysine	Arginine	G
G	Valine	Alanine	Aspartate	Glycine	U
	Valine	Alanine	Aspartate	Glycine	C
	Valine	Alanine	Glutamate	Glycine	A
	Valine	Alanine	Glutamate	Glycine	G

A codon consists of three nucleotides read in the sequence shown above. For example ACU codes threonine. The first letter, A, is read in the first column; the second letter, C, from the second letter columns; and the third letter, U, from the third letter column. Each of the codons is recognized by a corresponding anticodon sequence on a tRNA molecule. Some tRNA molecules recognize more than one codon sequence but always for the same amino acid. Most amino acids are encoded by more than one codon. For example, threonine is encoded by four codons (ACU, ACC, ACA, ACG), which differ from one another only in the third position.

1. Define the following terms:

   Entrons

   Transcription

   Exons

   Translation

2. Compare the structure and function of mRNA, tRNA, and rRNA molecules.

3. Compare a DNA molecule to an RNA molecule.

4. Explain the processes of transcription and translation as incorporated into recombinant DNA technology.

*Use all references and materials at your disposal to answer these review questions.

EPITHELIAL TISSUE
Simple epithelial tissue

13

OBJECTIVES

1 Identify cells and tissue associated with covering and lining simple epithelial tissue.

2 Draw and label covering and lining simple epithelial tissue.

3 Explain the location and function of covering and lining simple epithelial tissue.

KEY TERMS

Basement membrane
Epithelial tissue
Histology
Nucleus
Plasma membrane
Simple columnar epithelium
Simple cuboidal epithelium
Simple epithelium
Simple squamous epithelium
Tissue

MATERIALS

compound microscope
pencil
colored pencils

slides of the human body
system pertaining to simple
epithelial tissues

Similar cells that perform specialized activities and are grouped together form a **tissue.** Some tissues protect the body, some tissues produce chemicals, some tissues aid in digestion, and some tissues support and move the body. The study of tissue is called **histology.**

Epithelial tissue lines body cavities, covers other tissue and body surfaces, and forms glands.

Epithelial cells come in various shapes and provide protection, filtration, absorption, and secretion functions. Epithelium overlies and adheres to connective tissue via a basement membrane, which is a sticky, nonliving cellular secretion of the epithelium.

All epithelial tissues have a free surface. This means that one surface is exposed to the outside or to an internal air or fluid-filled cavity. Therefore epithelial tissues line body cavities as well as cover body surfaces including some internal organs. Also epithelial tissues are tightly interlocking; they have very little intercellular space. Epithelial tissues may be classified according to the shape of the cells and the number of layers of cells. The term **simple** means that there is one layer of cells. This arrangment is desirable in places where transport (through only one layer of cells) is a major function.

Simple squamous epithelium (Color Plate 3, *A*)
Simple cuboidal epithelium (Color Plate 3, *B*)
Simple columnar epithelium (Color Plate 3, *C*)

PROCEDURE A   SIMPLE EPITHELIAL TISSUE

1. Obtain the various prepared slides to identify cell shape. (Note the "purple dot," which generally indicates the **nucleus** of the cell; the purple line around the granular area indicates the **plasma membrane.**)
2. Draw and label simple squamous epithelium, simple cuboidal epithelium, nonciliated simple columnar epithelium, and ciliated simple columnar epithelium.
3. Provide the following information for each observation:

Simple squamous epithelium

Magnification:

Description:

Function:

Location:

Simple cuboidal epithelium

Magnification:

Description:

Function:

Location:

Simple columnar epithelium

Magnification:

Description:

Function:

Location:

1. Define the term *tissue*.

2. What is the function of epithelial covering and lining tissue?

3. What is the function of simple squamous epithelium?

4. Give three locations of simple squamous epithelium.

5. What is the function of simple cuboidal epithelium?

6. Give two locations of simple cuboidal epithelium.

7. Compare the function of ciliated and nonciliated simple columnar epithelium.

8. Give an example of a location of both ciliated and nonciliated simple columnar epithelium.

*Use all references and materials at your disposal to answer these review questions.

EPITHELIAL TISSUE
Stratified epithelium and pseudostratified epithelium

14

OBJECTIVES

1 Identify cells and tissue associated with stratified and pseudostratified epithelial tissue.
2 Draw and label stratified and pseudostratified epithelial tissue.
3 Explain the location and function of stratified and pseudostratified epithelial tissue.

KEY TERMS

Pseudostratified epithelium
Stratified columnar epithelium
Stratified epithelium
Stratified squamous
  epithelium

MATERIALS

compound microscope
pencil
colored pencils

slides of the human body
  system pertaining to stratified
  and pseudostratified
  epithelium

Cells arranged in several layers are called **stratified epithelial** cells. They provide protection for underlying tissue and may also produce secretions. **Stratified epithelium** is classified according to the cell shape:

  Stratified squamous epithelium (Color Plate 4, *A*)
  Stratified cuboidal epithelium (Color Plate 4, *B*)
  Stratified columnar epithelium (Color Plate 4, *C*)
  Pseudostratified eipthelium (Color Plate 4, *D*)

PROCEDURE A   STRATIFIED EPITHELIAL TISSUE

1. Obtain various prepared slides to identify, draw, and label stratified and pseudostratified epithelial tissue.

2. Provide the following information for each observation:

Stratified squamous epithelium

Magnification:

Description:

Function:

Location:

Stratified cuboidal epithelium

Magnification:

Description:

Function:

Location:

Stratified columnar epithelium

Magnification:

Description:

Function:

Location

Pseudostratified epithelium

Magnification:

Description:

Function:

Location:

1. Give two locations and explain the function of stratified squamous epithelium.

2. Where is stratified cuboidal epithelium located? What is the function?

3. Explain the function and give an example of a location of stratified columnar epithelium.

4. How is the structure of pseudostratified epithelium related to the function? Give an example.

*Use all references and materials at your disposal to answer these review questions.

EPITHELIAL TISSUE
Glandular epithelium

15

OBJECTIVES

1 Identify cells and tissue associated with glandular epithelial tissue.
2 Draw and label glandular epithelial tissue.
3 Explain the location and function of glandular epithelial tissue.

KEY TERMS

Apocrine gland
Branched acinar gland
Compound acinar gland
Compound tubular gland
Endocrine gland
Exocrine gland
Glandular epithelium
Goblet cell
Holocrine gland
Merocrine gland
Multicellular gland
Simple acinar
Simple coiled tubular gland
Simple tubular gland
Unicellular gland

MATERIALS

compound microscope
pencil
colored pencils

slides of the human body
system pertaining to
glandular epithelium

Embedded within the covering and lining epithelial tissue is **glandular epithelium.** The cell or groups of cells provide secretions. **Endocrine glands** secrete only hormones, which are released directly into the blood. **Exocrine** glands produce sweat, oil, and enzymes, which are released at the surface of the covering and lining epithelium (see Color Plates 9, B and D). Exocrine glands are classified according to their function.

An exocrine gland whose secretory product collects in the cytoplasm and is released upon death of the cell is called a **holocrine gland.** An exocrine gland whose secretory product is released from the cell is called a **merocrine gland.** In an **apocrine gland** the secretory product collects at the surface of the cell and pinches off.

Exocrine glands are also classified according to the shape of the secretory portion. **Unicellular glands** are one-celled glands that secrete mucus (Fig. 15-1). An example of a unicellular gland is a **goblet cell** (Color Plate 5). **Multicellular glands** are also classified according to the overall shape of the gland. They include simple straight tubular, simple coiled tubular, compound tubular, simple acinar, branched acinar, and compound acinar (Fig. 15-2).

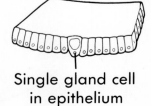

Single gland cell
in epithelium

FIG. 15-1 Unicellular gland.

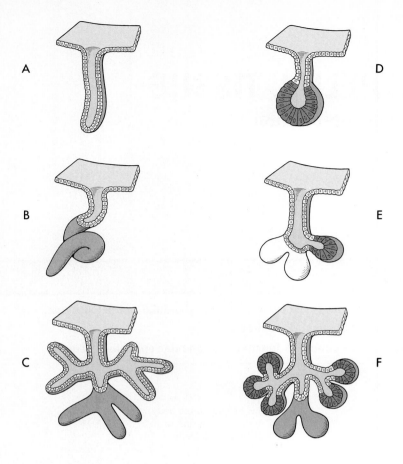

FIG. 15-2 Exocrine gland structures. **A,** Simple straight tubular. **B,** Simple coiled tubular. **C,** Compound tubular. **D,** Simple acinar. **E,** Branched acinar. **F,** Compound acinar.

A

D

B

E

C

F

PROCEDURE A   GLANDULAR EPITHELIAL TISSUE

1. Obtain prepared slides to identify, draw, and label glandular epithelial tissue types.
2. Provide the following information for each observation:

Unicellular exocrine glands

Magnification:

Description:

Function:

Location:

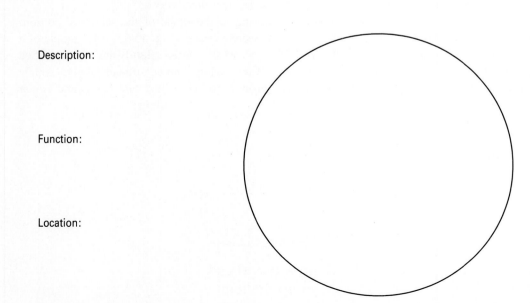

Simple tubular exocrine glands

Magnification:

Description:

Function:

Location:

Compound tubular exocrine glands

Magnification:

Description:

Function:

Location:

Simple coiled tubular exocrine glands

Magnification:

Description:

Function:

Location:

Simple acinar exocrine glands

Magnification:

Description:

Function:

Location:

Branched acinar exocrine glands

Magnification:

Description:

Function:

Location:

Compound acinar exocrine glands

Magnification:

Description:

Function:

Location:

1. What is a gland?

2. Compare the function of an endocrine gland and an exocrine gland.

3. Define the following terms:

   Holocrine gland

   Merocrine gland

   Apocrine gland

4. Describe the difference between a unicellular exocrine gland and a multicellular exocrine gland.

5. Give an example of a location and the function of the following:

   Goblet cell

   Simple tubular gland

   Simple coiled tubular gland

   Compound acinar gland

*Use all references and materials at your disposal to answer these review questions.

CONNECTIVE TISSUE
Adult connective tissue proper

16

OBJECTIVES
1 Identify cells and tissue associated with connective tissue proper.
2 Draw and label connective tissue proper.
3 Explain the location and function of connective tissue proper.

KEY TERMS

Adipose connective tissue
Adult connective tissue
Collagenous fibers
Connective tissue
Connective tissue proper
Dense connective tissue
Elastic connective tissue
Elastic fibers
Embryonic connective tissue
Fibroblast
Loose connective tissue
Matrix
Reticular connective tissue
Reticular fibers

MATERIALS

compound microscope
pencil
colored pencils

slides of the human body
system pertaining to adult
connective tissue proper

Connective tissue serves protection, support, and binding functions. The cells making up connective tissue are widely scattered and produce a nonliving substance called the **matrix (intercellular material).** The matrix may contain three types of fibers embedded between the cells. **Collagenous fibers,** made of the protein collagen, are resistant to tension yet somewhat flexible. **Elastic fibers,** made of the protein elastin, can stretch up to 50% of their length. **Reticular fibers** are short fibers made of immature collagen and carbohydrates, and provide support and strength.

Connective tissue is classified as **embryonic connective tissue** or **adult connective tissue.** One type of adult connective tissue is **connective tissue proper.** It has a fluidlike intercellular material, and the typical cell is a **fibroblast.** Adult connective tissue proper includes **loose connective tissue, dense connective tissue, elastic connective tissue, adipose tissue,** and **reticular connective tissue** (Color Plate 6).

PROCEDURE A   CONNECTIVE TISSUE PROPER

1. Obtain various prepared slides to identify, draw and label types of adult connective tissue proper.
2. Provide the following information for each observation:

Loose (areolar) connective tissue

Magnification:

Description:

Function:

Location:

Adipose connective tissue

Magnification:

Description:

Function:

Location:

Dense connective (collagenous) tissue

Magnification:

Description:

Function:

Location:

Elastic connective tissue

Magnification:

Description:

Function:

Location:

Reticular connective tissue

Magnification:

Description:

Function:

Location:

1. Compare connective tissue to epithelial tissue.

2. Define the following terms:

   Fibroblast

   Matrix

   Collagen

   Elastin

3. Give an example of a location and explain the function of loose connective tissue.

4. Why is adipose connective tissue important?

5. Give an example of a location and explain the function of dense connective tissue.

6. What type of connective tissue is found in the lungs and artery walls?

7. How is the function of reticular connective tissue related to the structure?

*Use all references and materials at your disposal to answer these review questions.

17

CONNECTIVE TISSUE
Cartilaginous connective tissue

OBJECTIVES

1 Identify cells and tissue associated with cartilaginous connective tissue.
2 Draw and label cartilaginous connective tissue.
3 Explain the location and function of cartilaginous connective tissue.

MATERIALS

compound microscope
pencil
colored pencils

slides of the human body
    system pertaining to carti-
    laginous connective tissue

The gel-like intercellular material of **cartilaginous connective tissue** provides strength and flexibility. The fibers within this matrix are **collagenous fibers** and **elastic fibers.** Mature cells within this matrix are called **chondrocytes,** which are housed in spaces called **lacunae.**

Three kinds of cartilaginous tissue are **hyaline cartilage, fibrocartilage,** and **elastic cartilage** (Color Plate 7).

PROCEDURE A   CARTILAGINOUS CONNECTIVE TISSUE

1. Obtain various prepared slides to identify, draw, and label types of cartilaginous tissues.
2. Provide the following information for each observation.
   Hyaline cartilage

Magnification:

Description:

Function:

Location:

Fibrocartilage

Magnification:

Description:

Function:

Location:

Elastic cartilage

Magnification:

Description:

Function:

Location:

1. Compare the matrix of adult connective tissue proper to cartilaginous connective tissue.

2. What cells are associated with cartilaginous connective tissue?

3. Give an example of a location and explain the function of hyaline cartilage.

4. Explain the function of fibrocartilage.

5. Where is fibrocartilage located?

6. Give an example of a location and explain the function of elastic cartilage.

*Use all references and materials at your disposal to answer these review questions.

ANATOMY AND HISTOLOGY OF A LONG BONE

18

OBJECTIVES

1 List and explain the anatomical features of a long bone.
2 List and describe the structures of compact bone tissue.
3 Compare the histological characteristics of spongy bone tissue to compact bone tissue.

MATERIALS

chicken leg bone (soaked overnight in water)
compact bone tissue slide
spongy bone tissue slide

pencil
colored pencils
compound microscope
dissecting microscope

KEY TERMS

Articular cartilage
Bone
Calcium carbonate
Calcium phosphate
Canaliculi
Cancellous bone
Cartilage
Collagenous fibers
Compact bone
Concentric ring
Connective tissue
Diaphysis
Endosteum
Epiphyseal line
Epiphyseal plate
Epiphysis
Haversian canal
Interstitial lamellae
Lacunae
Long bone
Matrix
Medullary cavity
Osseous
Osteoblast
Osteoclast
Osteocyte
Periosteum
Skeletal system
Trabeculae
Volkmann's canal

The **skeletal system** provides a structure of bones for muscle attachment. Bones act as levers. Muscles connect indirectly with bones via tendons. As muscles contract, tendons pull on bones and movement is produced. The skeletal system also protects internal organs, provides a storage area for mineral salts, and is a major site for blood cell production.

The skeletal system consists of the **connective tissues, cartilage,** and **bone. Osseous (bone) tissue,** like all connective tissue, has widely separated cells called **osteocytes,** with a considerable amount of **matrix.** The matrix primarily contains the mineral salts, **calcium phosphate,** and **calcium carbonate,** which provide bone its hardness. During bone development **collagenous fibers** combine with the mineral salts to provide reinforcement of the bone tissue.

The anatomy of a typical **long bone** consists of the following (Fig. 18-1):

Diaphysis   The long portion of the bone

Epiphysis   The end of the bone

Articular cartilage   The hyaline cartilage covering the epiphysis

Periosteum   Contains two layers: an **outer layer,** which contains blood vessels, lymph vessels, and nerves entering into the bone; and an **inner layer,** which contains blood vessels, elastic fiber, and **osteoblasts.** Osteoblasts provide bone growth and repair. The periosteum also provides attachment for muscles, tendons, and ligaments

Epiphyseal plate   Provides bone elongation as area of cartilaginous tissue is replaced by bone on the diaphysis side; about age 18-20 the cartilaginous cells stop dividing; this forms the **epiphyseal line**

Medullary cavity   Space in the long bone; contains the yellow marrow of adult bone

Endosteum   Lines the medullary cavity; contains osteoblasts and **osteoclasts.** Osteoclasts are involved in the removal of bone

FIG. 18-1 Anatomy of a typical long bone.

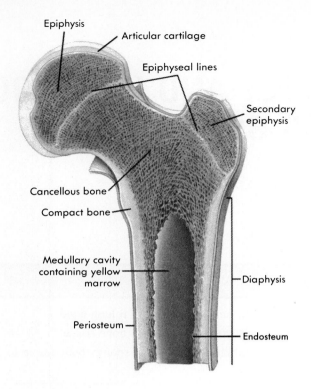

Epiphysis

Articular cartilage

Epiphyseal lines

Secondary epiphysis

Cancellous bone

Compact bone

Medullary cavity containing yellow marrow

Periosteum

Diaphysis

Endosteum

FIG. 18-2 Cancellous bone.

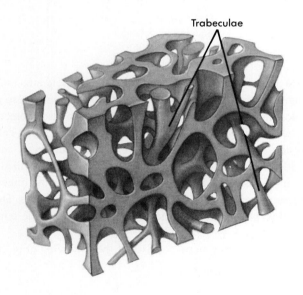

Trabeculae

FIG. 18-3   Compact bone.

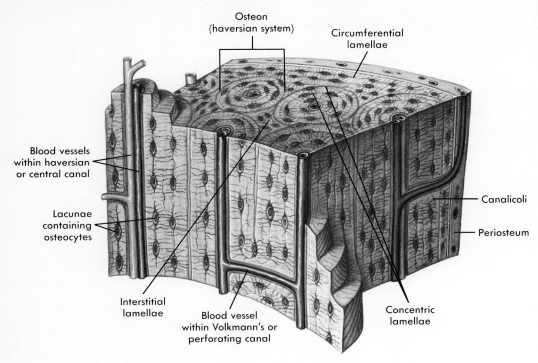

Osteon
(haversian system)

Circumferential
lamellae

Blood vessels
within haversian
or central canal

Lacunae
containing
osteocytes

Canalicoli

Periosteum

Interstitial
lamellae

Blood vessel
within Volkmann's or
perforating canal

Concentric
lamellae

Pores in bone tissue provide channels for blood vessels. If a great deal of porosity occurs, the bone is considered to be **cancellous bone** (Fig. 18-2 and Color Plate 8, *A*). When there is little porosity, the bone is called **compact bone** (Fig. 18-3 and Color Plate 8, *B*).

In compact bone, structures called **Volkmann's canals** allow passage of blood vessels and nerves perpendicular to the periosteum of a long bone. Volkmann's canals connect with longitudinal canals called an **osteon (haversian canals).** These canals also contain blood vessels. Around the canals bone cells called osteocytes and their calcified matrix appear as a series of **concentric rings,** which is called an **osteon (haversian system).** The matrix between two osteocytes is termed the interstitial **lamellae.** The **osteocytes** are housed in spaces called **lacunae.** The osteocytes communicate with one another by way of minute cytoplasmic canals called **canaliculi.**

Spongy bone appears as an irregular latticework called **trabeculae.** Within the trabeculae are scattered osteocytes within their lacunae. In some bones the large spaces between the trabeculae contain red marrow. In others, red marrow has been converted into yellow marrow. Cells within the red marrow produce new blood cells.

PROCEDURE A   LONG BONE ANATOMY

1. Obtain a chicken leg bone and identify and describe the diaphysis, epiphysis, articular cartilage, and periosteum.
2. Next, make a longitudinal cut along the bone. Now identify and describe the medullary cavity and the endosteum. Use the dissecting microscope to closely observe these same areas.
3. Draw, label, and describe the anatomical features of the long bone longitudinal section.
4. Continue using the dissecting microscope to compare spongy bone with compact bone.

PROCEDURE B COMPACT BONE HISTOLOGY

1. Using a compound microscope and a compact bone slide identify and describe the haversian canals, Volkmann's canals, interstitial lamellae, lacunae, osteocytes, and canaliculi.
2. Draw and label the above anatomical structures.
3. Provide the following information for each observation:

Magnification:

Description:

Location:

Function:

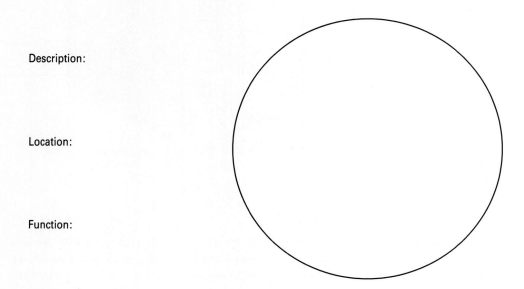

PROCEDURE C SPONGY BONE HISTOLOGY

1. Obtain a spongy bone slide to identify and describe the trabeculae, lacunae, osteocytes, and red marrow.
2. Draw and label the above anatomical structures.
3. Provide the following information for each observation:

Magnification:

Description:

Location:

Function:

1. Discuss how the structure of spongy bone relates to its function.

2. Explain how the structure of compact bone relates to its function.

3. Explain the function of the following long bone terms:

   Periosteum

   Articulating cartilage

   Endosteum

   Medullary cavity

   Haversian canal

   Volkmann's canal

4. Define the following terms:

   Osteoclast

   Osteoblast

   Osteocytes

   Lacunae

   Canaliculi

   Trabeculae

   Lamellae

*Use all references and materials at your disposal to answer these review questions.

19

CHEMICAL PROPERTIES OF BONE

KEY TERMS

Calcium carbonate
Calcium phosphate
Collagenous fibers
Hydroxyapatite
Inorganic salts
Organic collagen

OBJECTIVES

1 Compare the inorganic properties and organic properties of bone.

MATERIALS

3 chicken leg bones
beaker
metal tray

3% acetic acid
oven (150° C)
plastic bag

The interstitial chemical composition of bone consists of two main components: **inorganic salts** and **organic collagen.** The inorganic salts include **hydroxyapatite** crystals, **calcium phosphate,** and **calcium carbonate.** These salts comprise about 65% of the weight of bone and are responsible for its hardness. The **organic component** of bone, **collagenous fibers,** makes up about 35% of the weight of the bone and provides reinforcement for bone tissue. Removal of the inorganic salts or organic collagen demonstrates the unique chemical properties of bone (Fig. 19-1).

PROCEDURE A EXAMINATION OF THE CHEMICAL PROPERTIES OF BONE MATRIX

1. Place one chicken leg bone in a plastic bag to serve as a control.
2. Place the second chicken leg bone in a beaker and fill the beaker with 3% acetic acid. Allow the bone to soak for 24 hours.
3. Place the third chicken leg bone in a metal tray and place the tray in a 150° C oven for about 3 hours.
4. On the following day, compare properties of each bone.

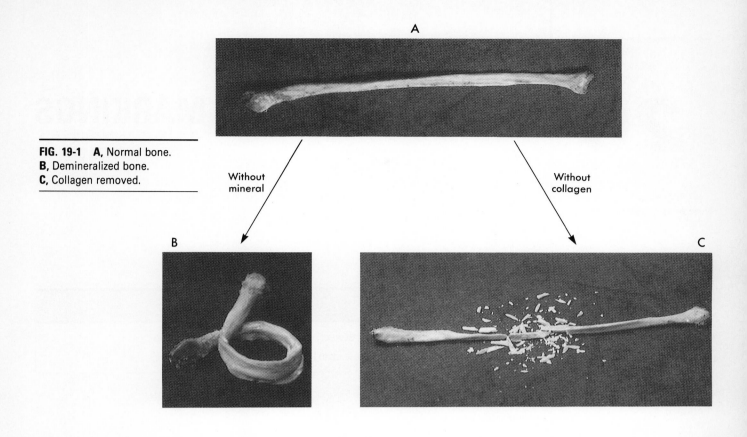

A

FIG. 19-1 **A,** Normal bone.
B, Demineralized bone.
C, Collagen removed.

Without
mineral

Without
collagen

B

C

1. Explain the importance of hydroxyapatite to bone tissue.

2. Discuss the importance of collagen fibers to bone tissue.

3. Define the term *ossification*.

4. Describe how intramembranous ossification occurs.

5. Explain how endochondral ossification occurs.

*Use all references and materials at your disposal to answer these review questions.

20

BONE MARKINGS

KEY TERMS

Condyle
Crest
Epicondyle
Facet
Fissure
Foramen
Fossa
Head
Line
Meatus
Process
Sinus
Spinous process
Sulcus
Trochanter
Tubercle
Tuberosity

OBJECTIVES

1 Explain the relationship between bone structure and function.
2 Relate the structure of bone markings to their function.

MATERIALS

human skeleton

Markings on bones can be used to identify the bone functions. Rounded ends provide sturdy joints. Bones with depressions receive these rounded ends. Rough areas on bone provide attachment for ligaments, tendons, and muscles. Pores through bones and grooves in the bone surfaces provide channels for blood vessels and nerves.

Markings that form projections are called **processes.**

Processes that form joints include the following:

Condyle   A large concave or convex articulating area (See Fig. 27-1, *A*)

Head   A rounded area on a constricted part of the bone (see Fig. 27-1, *A*)

Facet   A smooth, flat surface (see Fig. 24-6)

Processes that have muscles, tendons, and ligaments attached include:

Tubercle   A small rounded projection (See Fig. 25-3)

Tuberosity   A large rounded projection that is generally roughened (See Fig. 25-3)

Trochanter   A large blunt process (found only on the femur) (See Fig. 27-1, *A*)

Epicondyle   A projection above a condyle (see Fig. 27-1, *A*)

Spinous process   A sharp narrow projection (see Fig. 24-6)

Line   A small ridgelike projection (see Fig. 26-1, *B* and *C*)

Crest   A prominent ridgelike projection (see Fig. 26-1, *B* and *C*)

Markings that are openings or depressions include the following:

Foramen   A pore that provides a channel for blood vessels, nerves, or ligaments

Meatus   A tubelike channel (see Fig. 52-1)

Sulcus or groove   A groove that provides space for a blood vessel, nerve, or tendon (see Fig. 25-3)

Fossa   A depression on the bone (see Fig. 27-1, *A*)

Fissure   An open cavity between adjacent bones through which nerves or blood vessels pass

Sinus   Mucus-lined cavity

PROCEDURE A   BONE MARKING IDENTIFICATION

1. Identify and describe the following bone markings:

Condyle	Tuberosity	Line	Sulcus
Head	Trochanter	Crest	Fossa
Facet	Epicondyle	Foramen	Fissure
Tubercle	Spinous process	Meatus	

PROCEDURE B   BONE MARKING FUNCTIONS

1. Explain the function of each of the following bone markings:

Condyle

Head

Facet

Tubercle

Tuberosity

Trochanter

Epicondyle

Spinous process

Line

Crest

Foramen

Meatus

Sulcus

Fossa

Fissure

1. Give one location for each of the following bone markings:

   Meatus

   Condyle

   Head

   Facet

   Tubercle

   Tuberosity

   Crest

   Sinus

   Epicondyle

2. Explain how the structure in each of the following terms relates to its function:

   Meatus

   Condyle

   Facet

   Tubercle

   Tuberosity

   Crest

   Sinus

   Epicondyle

*Use all references and materials at your disposal to answer these review questions.

AXIAL SKELETAL SYSTEM
Cranial bones of the skull

21

OBJECTIVES	KEY TERMS

OBJECTIVES

1 Explain the relationship between bone structure and function.
2 Relate the structure of each marking to its function.
3 Identify the bones of the skull and the major markings associated with each.

MATERIALS

human skull

The **skull** contains cranial bones and facial bones. The **cranial bones** house and protect the brain and the organs of sight and hearing. The following bones and bone parts may be observed in various viewing positions.

Frontal bone	Forehead bone; also forms most of roof of **orbits** (eye sockets) and anterior part of floor
Supraorbital margin	Arched ridge just below eyebrow
Frontal sinuses	Cavities inside bone just above supraorbital margin; lined with mucosa and contain air
Frontal tuberosities	Bulge above each orbit; most prominent part of forehead
Superciliary arches	Ridges caused by projection of frontal sinuses; eyebrows lie over these ridges
Glabella	Smooth area between superciliary arches and above nose
Coronal suture	Line between frontal bone and two parietal bones
Parietal bones	Prominent, bulging bones behind frontal bone; form top sides of cranial cavity
Saggital suture	Line between parietal bones
Sphenoid bone	Keystone of cranial floor; forms its midportion; resembles bat wing outstretched; lies behind and slightly above nose and throat; forms part of floor and sidewalls of orbit
Body	Hollow, cubelike central portion
Greater wings	Lateral projections from body; form part of outer wall of orbit
Lesser wings	Thin, triangular projections from upper part of sphenoid body; form posterior part of roof of orbit
Sella turcica	Saddle-shaped depression on upper surface of sphenoid body; contains pituitary gland
Sphenoid sinuses	Irregular air-filled, mucosa-lined spaces within central part of sphenoid

KEY TERMS

Body
Condyles
Cranial bones
Cribriform plate
Crista galli
Ethmoid bone
Ethmoid sinuses
External auditory meatus
External occipital protuberance
Frontal bone
Frontal sinuses
Frontal tuberosities
Glabella
Greater wings
Inferior nuchal line
Internal occipital protuberance
Internal auditory meatus
Jugular fossa
Lateral masses
Lesser wings
Mandibular fossa
Mastoid air cells
Mastoid portion
Mastoid process
Occipital bone
Parietal bones
Perpendicular plate
Petrous portion
Pterygoid processes
Sella turcica
Skull
Sphenoid bone
Sphenoid sinuses
Squamous portion
Styloid process
Superciliary arches
Superior and middle turbinates
Superior nuchal line
Supraorbital margin
Temporal bones
Zygomatic process

Pterygoid processes	Downward projections on either side where body and greater wing unite; known as *feet of the bat.*
Superior orbital fissure	Slitlike opening into orbit; lateral to optic foramen; passageway for cranial nerves
Temporal bones	Forms lower sides of cranium and part of cranial floor; contains middle and inner ear structures
Mastoid process	Protuberance just behind ear; attachment site for some neck muscles
Mastoid air cells	Air-filled, mucosa-lined spaces
External auditory meatus	Opening into ear and tube extending into temporal bone
Zygomatic process	Projection that articulates zygomatic bone
Internal auditory meatus	Fairly large opening on posterior surface of petrous portion of bone; pathway for cranial nerve to inner ear and facial structures
Squamous portion	Thin, flaring upper part of bone
Petrous portion	Wedge-shaped process that forms part of center section of cranial floor between sphenoid and occipital bones
Mandibular fossa	Oval-shaped depression anterior to external auditory meatus; forms socket for condyle of mandible
Styloid process	Slender spike of bone extending downward and forward from undersurface of bone anterior to mastoid process; attachment site for some neck muscles; suspends hyoid bone via ligaments
Jugular fossa	Depression on undersurface of petrous portion
Squamosal suture	Between parietal bone and temporal bone
Occipital bone	Forms posterior part of cranial floor and walls
Condyles	Convex, oval processes on either side of **foramen magnum;** articulate with depressions on first cervical vertebra
External occipital protuberance	Prominent projection on posterior surface in midline short distance above foramen magnum; can be felt as definite bump
Superior nuchal line	Curved ridge extending laterally from external occipital protuberance
Inferior nuchal line	Less well-defined ridge paralleling superior nuchal line that is short distance below it
Internal occipital protuberance	Projection in midline on inner surface of bone; from this process, grooves for lateral sinuses extend laterally and one for sagittal sinus extends upward
Lamboidal suture	Line between parietal bones and occipital bone
Ethmoid bone	Complicated irregular bone that helps make up anterior portion of cranial floor, medial wall of orbits, upper parts of nasal septum and sidewalls, and part of nasal roof; lies anterior to sphenoid and posterior to nasal bones
Cribriform plate	Olfactory nerves pass through numerous holes in this plate
Crista galli	Process to which meninges attach
Perpendicular plate	Forms upper part of nasal septum
Ethmoid sinuses	Honeycombed, mucosa-lined air spaces within lateral masses of bone
Superior and middle turbinates or conchae	Delicate scroll-like plates of bone, which help form lateral walls of the nasal cavity

PROCEDURE A CRANIAL BONES

1. Locate each cranial bone and compare the structure to the function.
2. Locate the major markings of each cranial bone and relate the structure to the function.

1. Explain the function of the following cranial structures:

   Frontal sinus

   Sella turcica

   Petrous portion

   Styloid process

   Occipital condyle

   Crista galli

2. Define the paranasal sinsus.

3. What is sinusitis?

4. Name the cranial bones containing the paranasal sinuses.

*Use all references and materials at your disposal to answer these review questions.

22

AXIAL SKELETAL SYSTEM
Facial bones of the skull

OBJECTIVES

1 Explain the relationship between bone structure and function.
2 Relate the structure of each marking to its function.
3 Identify the facial bones and the major markings associated with each.

MATERIALS

human skull

The **skull** contains cranial bones and facial bones. The **facial bones** form the face and are all paired except the mandible and the vomer bones. The following bones and bone markings may be observed in various viewing positions:

Nasal bones	Small bones forming upper part of bridge of nose
Lacrimal bones	Thin bones about size and shape of fingernail; posterior and lateral to nasal bones in medial wall of orbit; help form sidewall of nasal cavity, a lacrimal duct; *often missing in dry skull*
Zygomatic bones	Cheekbones; form part of floor and sidewalls of orbit
Palatine bone	Forms posterior part of hard palate, floor, and part of walls of nasal cavity and floor of orbit
Horizontal plate	Joined to palatine processes of maxillae to complete part of hard palate
Mandible bone	Lower jawbone; largest, strongest, facial bone
Body	Main horizontal part of bone; forms chin
Ramus	Vertical process, one on either side, that projects upward from posterior part of body
Condyle	Part of ramus that articulates with mandibular fossa of temporal bone
Neck	Constricted part just below condyles
Alveolar process	Arch into which teeth are set
Coronoid process	Projection upward from anterior part of each ramus into which temporal muscle inserts
Angle	Juncture of the ramus and body

Maxilla bones	Upper jaw bones; form part of floor of orbit, anterior part of roof of mouth, floor of nose, and part of sidewalls of nose
Alveolar process	Arch containing teeth
Maxillary sinus	Large air-filled, mucosa-lined cavity within body of each maxilla; largest of all sinuses
Palatine process	Horizontal inward projections from alveolar process; forms anterior and larger part of hard palate
Lacrimal groove	Groove on inner surface; joined by similar groove on lacrimal bone to form canal housing nasolacrimal duct
Inferior conchae bones	Thin scroll of bone forming kind of shelf along surface of sidewall of nasal cavity; lies above roof of mouth
Vomer bone	Forms lower and posterior part of nasal septum
Ear bones (malleus, incus, stapes)	Tiny bones referred to as **auditory ossicles** in middle ear cavity in temporal bone

PROCEDURE A CRANIAL BONES

1. Use the skull model to locate and identify the facial bones.
2. Use the skull model to compare the facial bones and markings to their functions.

REVIEW QUESTIONS*

1. What is the smallest facial bone?

2. How does the structure of the inferior nasal conchae relate to their function?

3. Define cleft palate and cleft lip. Which bone is associated with these conditions?

4. How is the mandible bone different from other facial bones?

5. Which bones form the zygomatic arch?

6. What bones form the nasal septum?

7. Describe a dislocated jaw.

*Use all references and materials at your disposal to answer these review questions.

23

AXIAL SKELETAL SYSTEM
Foramina of the skull

OBJECTIVES

1 Identify and explain the functions of the principal foramina of the skull.

MATERIALS

human skull

Foramina of the skull are holes through which blood vessels, nerves, and ligaments pass. They may be classified according to their location and the structures they transmit. The following foramina may be observed in various viewing positions.

FORAMEN	LOCATION	STRUCTURE PASSING THROUGH
Carotid	Petrous portion of temporal bone	Internal carotid artery
Greater palatine	Posterior angle of hard palate	Greater palatine nerve and descending palatine vessels
Hypoglossal	Superior to base of occipital condyles	Hypoglossal nerve (XII) and branch of ascending pharyngeal artery
Incisive	Posterior to incisor teeth	Branches of descending palatine vessels and nasopalatine nerve
Intraorbital	In maxilla inferior to orbit	Infraorbital nerve and artery and maxillary branch of trigeminal nerve
Jugular	Posterior to carotid canal between petrous portion of temporal and occipital bones	Internal jugular vein, glossopharyngeal nerve (IX), vagus nerve (X), and accessory nerve (X)
Lacerum	Bounded anteriorly by sphenoid, posteriorly by petrous portion of temporal, and medially by sphenoid and occipital bones	Internal carotid artery and branch of ascending pharyngeal artery
Lacrimal	Lacrimal bone	Lacrimal (tear) duct
Lesser palatine	Posterior to greater palatine foramen	Lesser palatine nerves
Magnum	Occipital bone	Medulla oblongata and its membranes, accessory nerve (XI), and vertebral and spinal arteries
Mandibular	Medial surface of ramus of mandible	Inferior alveolar nerve and vessels and mandibular branch of trigeminal

FORAMEN	LOCATION	STRUCTURE PASSING THROUGH
Mastoid	Posterior border of mastoid process of temporal bone	Vein to transverse sinus and branch of occipital artery to dura mater
Mental	Inferior to second premolar tooth in mandible	Mental nerve and vessels
Olfactory	Cribriform plate of ethmoid bone	Olfactory nerve (I)
Optic	Between upper and lower portions of lesser wing of sphenoid bone	Optic nerve (III) and ophthalmic artery
Ovale	Greater wing of sphenoid bone	Mandibular branch of trigeminal nerve (V)
Rotundum	Junction of anterior and medial parts of sphenoid bone	Maxillary branch of trigeminal nerve (V)
Spinosum	Posterior angle of sphenoid bone	Middle meningeal vessels
Stylomastoid	Between styloid and mastoid processes of temporal bone	Facial nerve (VII) and stylomastoid artery
Supraorbital	Supraorbital margin of orbit	Supraorbital nerve and artery
Zygomaticofacial	Zygomatic bone	Zygomaticofacial nerve and vessels

PROCEDURE A   SKULL FORAMINA

1. Use the preceding list to locate and identify the foramina of the skull.
2. Name the structures passing through each of the skull foramina.

REVIEW QUESTIONS*

1. Match the following foramina with the proper bones or bone parts:

FORAMEN	BONE/BONE PART
Spinosum	Occipital
Stylomastoid	Cribriform plate
Ovale	Sphenoid
Optic	Mastoid
Olfactory	Petrous portion
Mastoid	Mastoid process
Carotid	Styloid process
Jugular	Lesser wing
Lacerum	Temporal
Magnum	Greater wing

*Use all references and materials at your disposal to answer these review questions.

24 AXIAL SKELETAL SYSTEM
Hyoid bone, spinal column, sternum, and ribs

OBJECTIVES

1 Identify the anatomical structures of the vertebrae.
2 Explain how the structure of the vertebrae relate to their function.
3 Identify and compare the structure and function of the sternum.
4 Describe the ribs and the relationship of their structure to their function.

MATERIALS

hyoid bone
thoracic vertebrae
sacrum
sternum
false ribs

cervical vertebrae
lumbar vertebrae
coccyx
true ribs

The axial skeletal system contains the **skull** as well as the **hyoid bone, vertebral column, sternum,** and **ribs.** The following bones and bone markings may be observed in various positions:

Hyoid bone (Fig. 24-1)	U-shaped bone in neck between mandible and upper part of larynx; only bone in body not forming joint with any other bone; suspended by ligaments from styloid processes of temporal bones
Spinal column (Fig. 24-2)	Flexible segmented rod shaped like elongated letter S; includes **cervical, thoracic, lumbar,** and **sacral curves;** forms axis of body; head balanced above, ribs and viscera suspended in front, and lower extremities attached below; encloses spinal cord
Cervical vertebrae	First or upper seven vertebrae
General features	Foramen in each transverse process for transmission of vertebral artery, vein, plexus of nerves; short bifurcated spinous processes except on seventh vertebrae, where it is extra long and may be felt as protrusion when head is bent forward; bodies of these vertebrae are small, whereas spinal foramina are large and triangular
Atlas (Fig. 24-3)	First cervical vertebra; lacks body and spinous process; superior articulating processes are concave ovals that act as socketlike cradles for condyles of occipital bone; named atlas because it supports head as mythical figure Atlas was thought to have supported the world; "yes" bone
Axis (Fig. 24-4)	Second cervical vertebra, so named because atlas rotates about this bone in rotating movements of head
Dens	Peglike projection upward from body of axis, forming pivot for rotation of atlas; "no" bone

FIG. 24-1 Hyoid bone.
A, Anterior view. **B,** Lateral view.

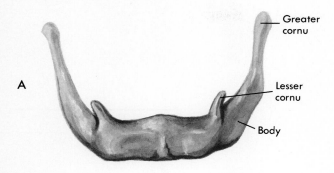

Greater
cornu

A

Lesser
cornu

Body

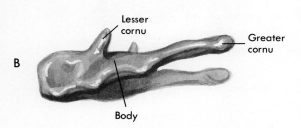

Lesser
cornu

Greater
cornu

B

Body

FIG. 24-2 Vertebral column.
Complete column viewed from
the left side.

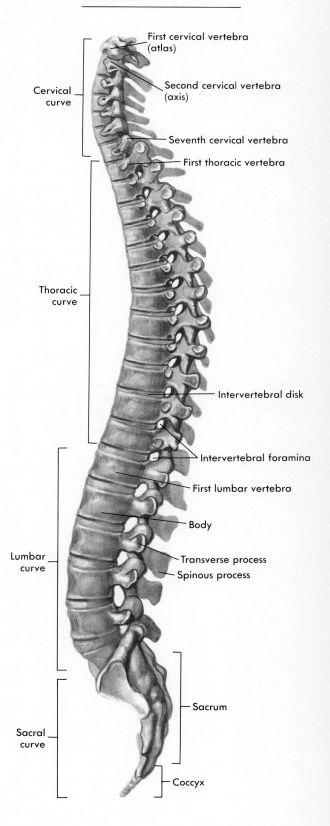

First cervical vertebra
(atlas)

Second cervical vertebra
(axis)

Cervical
curve

Seventh cervical vertebra

First thoracic vertebra

Thoracic
curve

Intervertebral disk

Intervertebral foramina

First lumbar vertebra

Body

Transverse process

Spinous process

Lumbar
curve

Sacrum

Sacral
curve

Coccyx

FIG. 24-3 Atlas.

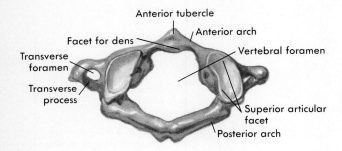

Anterior tubercle
Anterior arch
Facet for dens
Vertebral foramen
Transverse foramen
Transverse process
Superior articular facet
Posterior arch

FIG. 24-4 Axis.

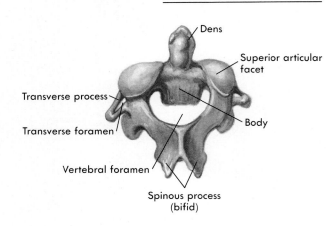

Dens
Superior articular facet
Transverse process
Body
Transverse foramen
Vertebral foramen
Spinous process (bifid)

FIG. 24-5 Thoracic vertebrae.

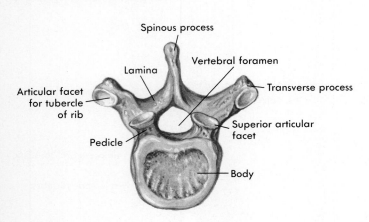

Spinous process
Vertebral foramen
Lamina
Transverse process
Articular facet for tubercle of rib
Superior articular facet
Pedicle
Body

Thoracic vertebrae (Fig. 24-5)	Next twelve vertebrae, to which twelve pairs of ribs are attached
Body	Main part; flat, round mass located anteriorly; supporting or weight-bearing part of vertebra
Pedicles	Short projections extending posteriorly from body
Laminae	Posterior part of vertebra to which pedicles join and from which processes project
Neural arch	Formed by pedicles and laminae; protects spinal cord posteriorly; congenital absence of one or more neural arches is known as **spina bifida**
Spinous process	Sharp process projecting inferiorly from laminae in midline
Transverse process	Right and left lateral projections of laminae
Superior articulating process	Project upward from laminae
Inferior articulating process	Project downward from laminae; articulate with superior articulating processes of vertebrae below
Vertebral or spinal foramen	Hole in center of vertebra formed by union of body, pedicles, and laminae; spinal foramina, when vertebrae are superimposed one on other, form spinal cavity that houses spinal cord

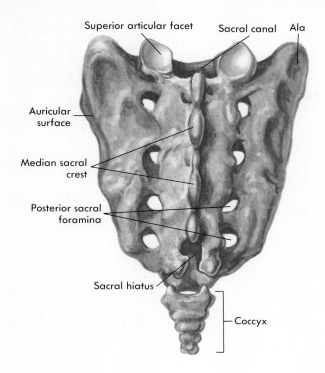

FIG. 24-6 Sacrum and coccyx.

Superior articular facet   Sacral canal   Ala

Auricular
surface

Median sacral
crest

Posterior sacral
foramina

Sacral hiatus

Coccyx

Lumbar vertebrae	Next five vertebrae; strong, massive, superior articulating processes directed inward instead of upward; inferior articulating processes, outward instead of downward; short, blunt spinous process
Sacrum (Fig. 24-6)	Five separate vertebrae until person is about 25 years of age, then fused to form wedged-shaped bone
Coccyx (Fig. 24-6)	Four or five separate vertebrae in child but fused into one in adult
Sternum and ribs (Fig. 24-7)	Sternum, ribs, and thoracic vertebrae together form bony cage known as **thorax;** ribs attach posteriorly to vertebrae and slant downward anteriorly to attach to sternum
Sternum (Fig. 24-7)	Breastbone; flat, dagger-shaped bone
Body	Main central part of bone
Manubrium	Flaring, upper part of bone
Xiphoid	Inferior border of bone
True ribs	Upper seven pairs; fasten directly to sternum by costal cartilages
False ribs	Do not attach to sternum directly; upper three pairs of false ribs attach by means of costal cartilage of seventh ribs; last two pairs do not attach to sternum at all, therefore called **floating ribs**
Head	Projection at posterior end of rib; articulates with corresponding thoracic vertebra and one above, except last three pairs, which join corresponding vertebrae only
Neck	Constricted portion just below head
Tubercle	Small knob just below neck; articulates with transverse process of corresponding thoracic vertebra; missing in lowest three ribs
Body	Main part of rib
Costal cartilage	Hyaline cartilage at **sternal end** of true ribs; attaches ribs (except floating ribs) to sternum

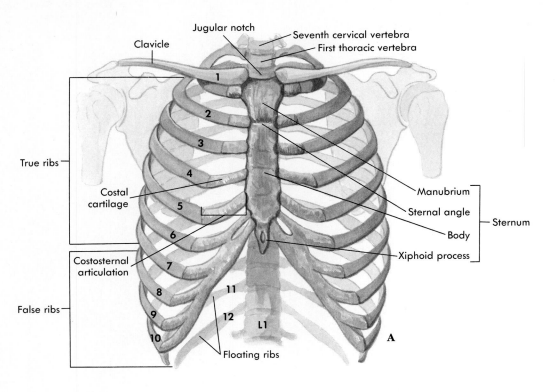

Jugular notch
Clavicle
Seventh cervical vertebra
First thoracic vertebra
1
2
3
True ribs
4
Costal cartilage
5
6
Costosternal articulation
7
8
False ribs
9
10
11
12
L1
Floating ribs
Manubrium
Sternal angle
Body
Sternum
Xiphoid process
A

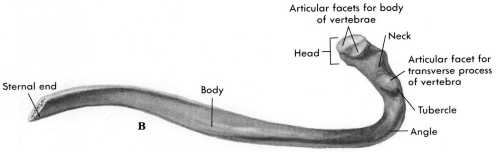

Articular facets for body of vertebrae
Head
Neck
Articular facet for transverse process of vertebra
Tubercle
Angle
Sternal end
Body
B

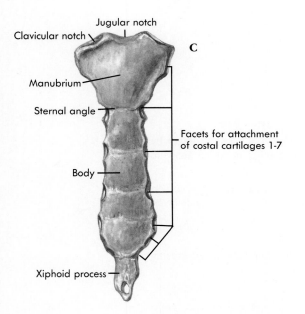

Jugular notch
Clavicular notch
C
Manubrium
Sternal angle
Body
Facets for attachment of costal cartilages 1-7
Xiphoid process

FIG. 24-7 A, Rib cage. **B,** Typical rib. **C,** Sternum.

PROCEDURE A   VERTEBRAL COLUMN

1. Use the human skeleton to locate and identify the cervical vertebrae, thoracic vertebrae, lumbar vertebrae, sacrum, and coccyx.
2. Draw and label a typical vertebra.
3. Explain the function of the vertebra in relationship to its structure.

PROCEDURE B   STERNUM

1. Draw and label the sternum.
2. Explain the function of the sternum in relation to its structure.

PROCEDURE C   RIBS

1. Use the human skeleton to locate and identify the true ribs, false ribs, and floating ribs.
2. Draw and label a typical rib.

1. Why is the hyoid bone unique?

2. What is the function of the hyoid bone?

3. Name the four curves of the spinal column.

4. Explain the development of the vertebral curves.

5. What is the function of the vertebral atlas?

6. What is the cause of fatality in whiplash injuries?

7. Explain the following conditions:

   Scoliosis
   Kyphosis
   Spina bifida

8. Describe a dislocated rib.

9. Compare the terms *true ribs*, *false ribs*, and *floating ribs*.

10. What is the function of the xiphoid process of the sternum?

*Use all references and materials at your disposal to answer these review questions.

APPENDICULAR SKELETAL SYSTEM
Pectoral Girdle and Upper Extremities

25

MATERIALS

human skeleton

The **appendicular skeletal system** includes the bones of the free **appendages (upper extremities** and **lower extremities)** and the **girdles (pectoral girdles** and **pelvic girdles),** which connect the free appendages to the axial skeleton.

Pectoral girdle	Scapulae and clavicles together make up shoulder girdle
CLAVICLE (Fig. 25-1)	Collar bones; shoulder girdle joined to axial skeleton by articulation of clavicles with sternum
SCAPULA (Fig. 25-2)	Shoulder blades; scapula does not form joint with axial skeleton
Superior border	Upper margin
Vertebral border	Margin toward vertebral column
Axillary border	Lateral margin
Spine	Sharp ridge running diagonally across posterior surface of shoulder blade
Acromion process	Slightly flaring projection at lateral end of scapular spine; may be felt as tip of shoulder; articulates with clavicle
Coracoid process	Projection on anterior surface from upper border of bone; may be felt in groove between deltoid and pectoralis major muscles
Glenoid cavity	Arm socket; receives head of humerus
Upper extremities	Articulates proximally with scapula and distally with ulna and radius
HUMERUS (Fig. 25-3)	
Head	Smooth, hemispherical enlargement at proximal end of humerus
Anatomical neck	Oblique groove just below head
Greater tubercle	Rounded projection lateral to head on anterior surface
Lesser tubercle	Prominent projection on anterior surface just below anatomical neck
Intertubercular	Deep groove between greater and lesser tubercles in which long tendon of biceps muscle lodges
Surgical neck	Region just below tubercles; so named because of its liability to fracture

KEY TERMS

Acromion process
Anatomical neck
Appendages
Appendicular skeletal system
Axillary border
Capitate
Capitulum
Carpals
Clavicle
Coracoid process
Coronoid fossa
Coronoid process
Deltoid tuberosity
Glenoid cavity
Greater tubercle
Hamate
Head
Humerus
Intertubercular
Lateral epicondyle
Lesser tubercle
Lunate
Medial epicondyle
Metacarpals
Olecranon fossa
Olecranon process
Phalanges
Pisiform
Pollex
Radial groove
Radial tuberosity
Radius
Scaphoid
Scapula
Semilunar notch
Shoulder girdle
Spine
Styloid process
Superior border
Surgical neck
Trapezium
Trapezoid
Triquetrum

Continued.

KEY TERMS

Trochlea
Ulna
Upper extremities
Vertebral border

FIG. 25-1 Clavicle.

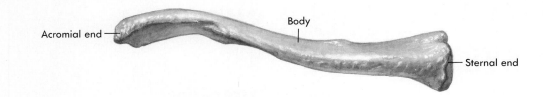

Acromial end ———

——— Body

——— Sternal end

FIG. 25-2 Scapula (left).
A, Anterior view.
B, Posterior view.

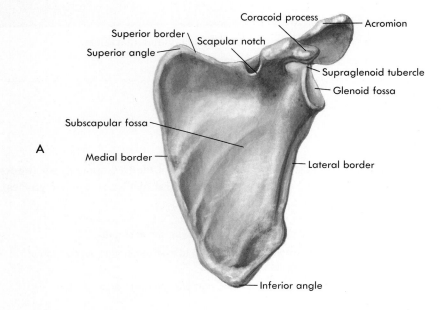

Coracoid process
Acromion
Superior border
Scapular notch
Superior angle
Supraglenoid tubercle
Glenoid fossa
Subscapular fossa
Medial border
Lateral border
Inferior angle

A

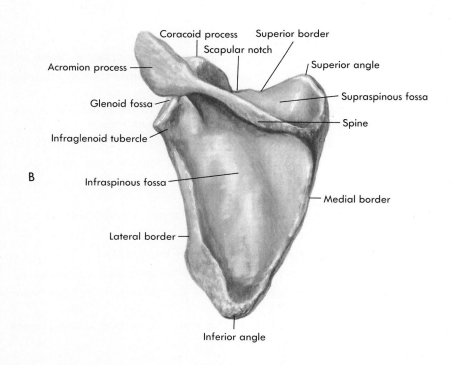

Coracoid process   Superior border
Scapular notch
Acromion process
Superior angle
Supraspinous fossa
Glenoid fossa
Spine
Infraglenoid tubercle
Infraspinous fossa
Medial border
Lateral border
Inferior angle

B

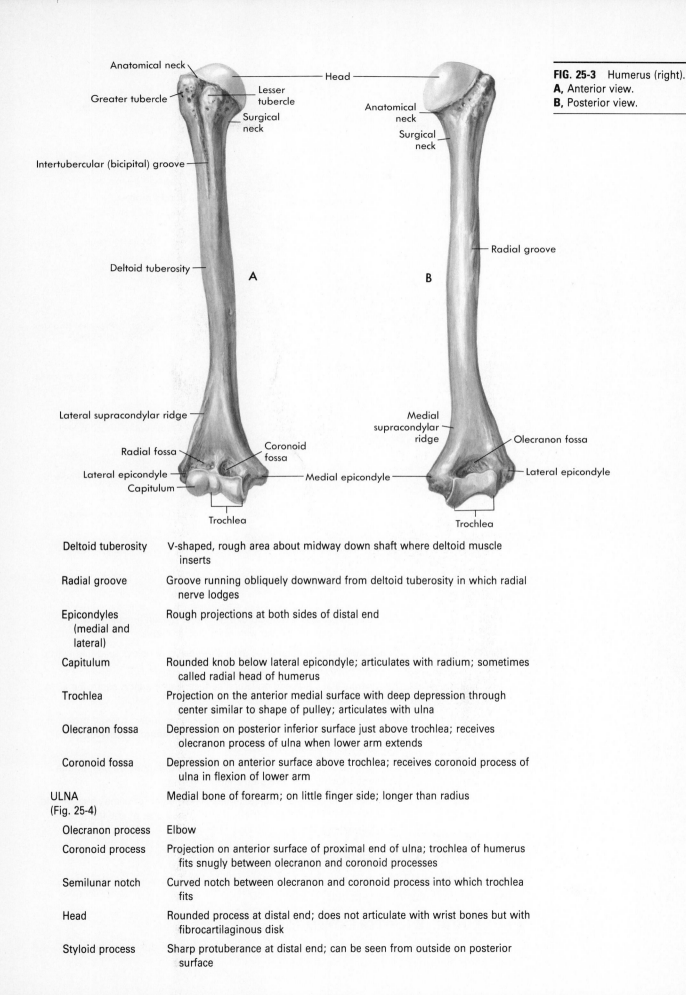

Anatomical neck

Greater tubercle

Head

Lesser tubercle

Surgical neck

Anatomical neck

Surgical neck

Intertubercular (bicipital) groove

Radial groove

Deltoid tuberosity

A

B

Lateral supracondylar ridge

Medial supracondylar ridge

Olecranon fossa

Radial fossa

Coronoid fossa

Lateral epicondyle

Lateral epicondyle

Medial epicondyle

Lateral epicondyle

Capitulum

Trochlea

Trochlea

FIG. 25-3 Humerus (right).
A, Anterior view.
B, Posterior view.

Deltoid tuberosity	V-shaped, rough area about midway down shaft where deltoid muscle inserts
Radial groove	Groove running obliquely downward from deltoid tuberosity in which radial nerve lodges
Epicondyles (medial and lateral)	Rough projections at both sides of distal end
Capitulum	Rounded knob below lateral epicondyle; articulates with radium; sometimes called radial head of humerus
Trochlea	Projection on the anterior medial surface with deep depression through center similar to shape of pulley; articulates with ulna
Olecranon fossa	Depression on posterior inferior surface just above trochlea; receives olecranon process of ulna when lower arm extends
Coronoid fossa	Depression on anterior surface above trochlea; receives coronoid process of ulna in flexion of lower arm
ULNA (Fig. 25-4)	Medial bone of forearm; on little finger side; longer than radius
Olecranon process	Elbow
Coronoid process	Projection on anterior surface of proximal end of ulna; trochlea of humerus fits snugly between olecranon and coronoid processes
Semilunar notch	Curved notch between olecranon and coronoid process into which trochlea fits
Head	Rounded process at distal end; does not articulate with wrist bones but with fibrocartilaginous disk
Styloid process	Sharp protuberance at distal end; can be seen from outside on posterior surface

FIG. 25-4 Ulna and radius (right).

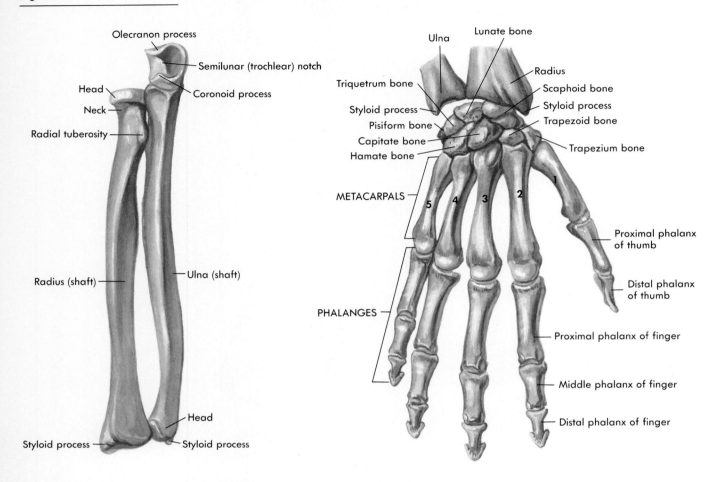

RADIUS (Fig. 25-4)	Lateral bone of forearm; on thumb side
Head	Disk-shaped process forming proximal end of radius; articulates with capitulum of humerus and with radial notch of ulna
Radial tuberosity	Roughened projection on ulnar side, short distance below head into which biceps muscle inserts
Styloid process	Protuberance at distal end on lateral surface (with forearm supinated as in anatomical position)
CARPALS (Fig. 25-5)	Eight separate bones forming the wrist; arranged in two rows at proximal end of hand (**scaphoid, lunate, triquetrum pisiform, trapezium, trapezoid, capitate,** and **hamate**)
METACARPALS (Fig. 25-5)	Long bones forming framework of palm of hand; numbered one, on the thumb side, through five on the little finger side
PHALANGES (Fig. 25-5)	Miniature long bones of fingers, three in each finger (proximal, medial, and distal), two in each thumb (**pollex);** numbered one, on the thumb side, through number five on the little finger side

PROCEDURE A   SHOULDER GIRDLE

1. Use the individual bones and the model skeleton to identify and describe the bones and bone markings of the shoulder girdle.

PROCEDURE B   UPPER EXTREMITIES

1. Use the individual bones and the model skeleton to identify and describe the bones and bone markings of the upper extremities.

REVIEW QUESTIONS*

1. Which bone in the body is the most frequently broken? Why?

2. What joint is involved in separation of the shoulder?

3. What is the function of the pectoral girdle?

4. Explain the function of each of the following appendicular skeletal markings:

    Coronoid tubercle

    Costal tuberosity

    Spine of scapula

    Supraspinatous fossa

    Deltoid tuberosity

    Radial tuberosity

    Olecranon fossa

5. A broken wrist usually involves which carpal bone?

*Use all references and materials at your disposal to answer these review questions.

26 APPENDICULAR SKELETAL SYSTEM
Pelvic girdle

OBJECTIVES

1 Identify the bones of the pelvic girdles and their major markings.
2 Compare the structural differences between the male and female pelvis.

MATERIALS

human skeleton

The **appendicular skeletal system** includes the bones of the **free appendages (upper extremities** and **lower extremities)** and the **girdles (pectoral girdles** and **pelvic girdles)** which connect the free appendages to the axial skeleton.

Pelvic Girdle (Fig. 26-1)	Large hip bones; with sacrum and coccyx, these three bones form basinlike pelvic cavity; lower extremities attached to axial skeleton by pelvic bones
Os Coxa	Formed by fusion of ilium, ischium, and pubis during development
Ilium	Upper, flaring portion
Ischium	Lower, posterior portion
Pubis	Medial, anterior section
Acetabulum	Hip socket; formed by union of ilium, ischium, and pubis
Iliac crests	Upper, curving boundary of ilium
Anterior superior spine	Prominent projection at anterior end of iliac crest; can be felt externally as "point" of hip
Anterior inferior spine	Less prominent projection short distance below anterior superior spine
Posterior superior spine	At posterior end of iliac crest
Posterior inferior spine	Just below posterior superior spine
Greater sciatic notch	Large notch on posterior surface of ilium just below posterior inferior spine
Gluteal lines	Three curved lines across outer surface of ilium—posterior, anterior, inferior, respectively
Iliopectineal line	Rounded ridge extending from pubic tubercle upward and backward toward sacrum
Iliac fossa	Large, smooth concave inner surface of ilium above iliopectineal line
Ischial tuberosity	Large, rough, quadrilateral process forming inferior part of ischium; in erect sitting position, body rests on these tuberosities

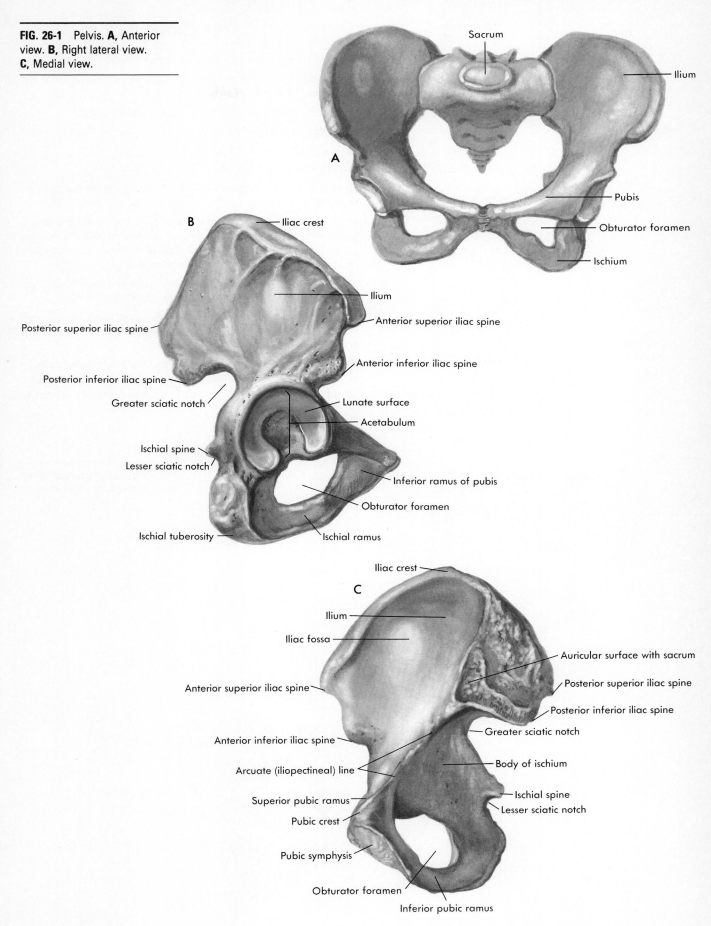

FIG. 26-1 Pelvis. **A,** Anterior view. **B,** Right lateral view. **C,** Medial view.

A

Sacrum

Ilium

Pubis

Obturator foramen

Ischium

B

Iliac crest

Ilium

Anterior superior iliac spine

Posterior superior iliac spine

Anterior inferior iliac spine

Posterior inferior iliac spine

Lunate surface

Greater sciatic notch

Acetabulum

Ischial spine

Lesser sciatic notch

Inferior ramus of pubis

Obturator foramen

Ischial tuberosity

Ischial ramus

C

Iliac crest

Ilium

Iliac fossa

Auricular surface with sacrum

Posterior superior iliac spine

Anterior superior iliac spine

Posterior inferior iliac spine

Greater sciatic notch

Anterior inferior iliac spine

Body of ischium

Arcuate (iliopectineal) line

Ischial spine

Superior pubic ramus

Lesser sciatic notch

Pubic crest

Pubic symphysis

Obturator foramen

Inferior pubic ramus

Ischial spine	Pointed projection just above tuberosity; landmark in measuring width of pelvic cavity
Symphysis pubis	Cartilaginous, amphiarthrotic joint between pubic bones
Superior pubic ramus	Part of pubis lying between symphysis and acetabulum; forms upper part of obturator foramen
Inferior pubic ramus	Part extending down from symphysis; unites with ischium
Pubic arch	Angle formed by two inferior rami
Pubic crest	Upper margin of superior ramus
Pubic tubercle	Rounded process at end of crest
Obturator foramen	Large hole in anterior surface of os coxa; formed by pubis and ischium; largest foramen in body
Pubic inlet	Boundary of aperture leading into true pelvis; formed by pubic crests, iliopectineal lines, and sacral promontory; size and shape of this inlet has great obstetrical importance, because if any of its diameters are too small, infant skull cannot enter true pelvis for natural birth
True pelvis	Space below pelvic brim; true "basin" with bone and muscle walls and muscle floor; pelvic organs located in this space
False pelvis	Broad, shallow space above pelvic brim, or pelvic inlet; name "false pelvis" is misleading, because this space is actually part of abdominal cavity, not pelvic cavity
Pelvic outlet	Irregular circumference marking lower limits of true pelvis; bounded by tip of coccyx and two ischial tuberosities

Male and female pelvic girdles are slightly different. The difference is primarily related to pregnancy and childbirth (Fig. 26-2 and Table 26-1).

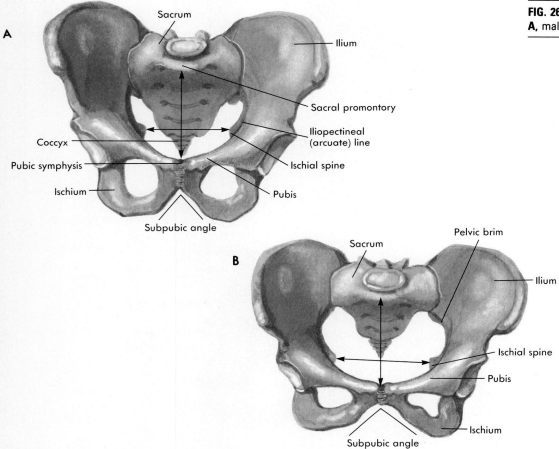

FIG. 26-2 Comparison between A, male and B, female pelvis.

A

Sacrum
Ilium
Sacral promontory
Iliopectineal (arcuate) line
Coccyx
Ischial spine
Pubic symphysis
Pubis
Ischium
Subpubic angle

B

Pelvic brim
Sacrum
Ilium
Ischial spine
Pubis
Ischium
Subpubic angle

TABLE 26-1 Comparison between male and female pelvis

POINT OF COMPARISON	MALE	FEMALE
General structure	Heavy and thick	Light and thin
Joint surfaces	Large	Small
Muscle attachments	Well marked	Indistinct
Pelvic inlet	Heart-shaped	Larger and oval
Sacrum	Long and narrow	Short and wide
Pubic arch	Less than 90 degrees	Greater than 90 degrees
Acetabulum	Large	Small
Greater sciatic notch	Narrow	Wide

PROCEDURE A   PELVIC GIRDLE

1. Use the individual bones and the model skeleton to identify and describe the bones and bone marking of the pelvic girdle.

PROCEDURE B   MALE AND FEMALE PELVIC GIRDLE COMPARISON

1. Use male and female skeletons to compare the structural differences in the pelvic girdles.

REVIEW QUESTIONS*

1. What is the function of the pelvic girdle?

2. What is the function of the following appendicular skeletal markings?

   Anterior superior iliac spine

   Acetabulum

3. Explain why the coxal bone is divided into three different areas.

4. Compare the male and female pelvic girdles.

*Use all references and materials at your disposal to answer these review questions.

27

APPENDICULAR SKELETAL SYSTEM
Lower extremities

KEY TERMS

Adductor
Appendicular skeletal system
Calcaneus
Condyle
Crest
Cuboid
Femur
Fibula
First cuneiform
Gluteal tubercle
Greater trochanter
Hallux
Head
Intercondylar eminence
Intercondyloid notch
Lateral malleolus
Lesser trochanter
Linea aspera
Lower extremities
Medial malleolus
Metatarsals
Navicular
Neck
Patella
Phalanges
Popliteal line
Second cuneiform
Supracondylar ridge
Talus
Tarsals
Third cuneiform
Tibia
Tibial tuberosity
Trochlea

OBJECTIVES

1 Identify the lower extremity bones and their markings.
2 Define the bones of the foot and explain the importance of the arches of the foot.

MATERIALS

human skeleton

The **appendicular skeletal system** includes the bones of the **free appendages (upper extremities** and **lower extremities)** and the **girdles (shoulder girdles** and **hip girdles)**, which connect the free appendages to the axial skeleton.

Femur (Fig. 27-1)	Thigh bone; largest, strongest bone of the body
Head	Rounded, upper end of bone; fits into acetabulum
Neck	Constricted portion just below head
Greater trochanter	Protuberance located inferiorly and laterally to head
Lesser trochanter	Small protuberance located inferior and medial to greater trochanter
Linea aspera	Prominent ridge extending lengthwise along concave posterior or surface
Gluteal tubercle	Rounded projection just below greater trochanter; rudimentary third trochanter
Gluteal tuberosity	Roughened area around the linea aspera; attachment site for large gluteal muscles
Supracondylar ridges	Two ridges formed by division of linea aspera at its lower end; medial supracondylar ridge extends inward to inner condyle, lateral ridge to outer condyle
Condyles	Large, rounded bulges at distal end of femur; one on medial and one on lateral surface
Adductor	Small projection just above inner condyle; marks termination of medial supracondylar ridge
Trochlea	Smooth depression between condyles on anterior surface; articulates with patella
Intercondyloid notch	Deep depression between condyles on posterior surface; cruciate ligaments that help bind femur to tibia lodge in this notch
Patella	Kneecap; largest sesamoid bone (embedded in tendon) of body

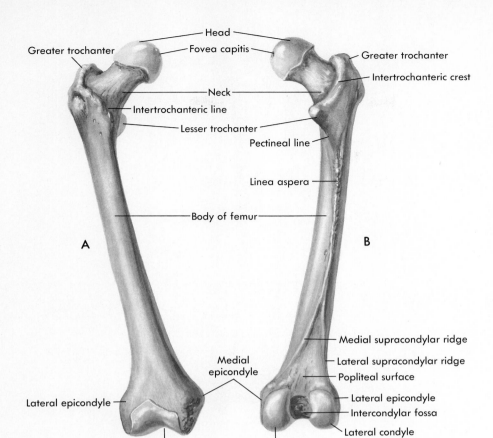

FIG. 27-1 Femur (right).
A, Anterior view. **B**, Posterior view.

Head
Greater trochanter — Fovea capitis
Greater trochanter
Intertrochanteric crest
Neck
Intertrochanteric line
Lesser trochanter
Pectineal line
Linea aspera
Body of femur
A
B
Medial epicondyle
Medial supracondylar ridge
Lateral supracondylar ridge
Popliteal surface
Lateral epicondyle
Intercondylar fossa
Lateral condyle
Lateral epicondyle
Patellar groove
Medial condyle

Tibia (Fig. 27-2)	Shinbone
Condyles	Bulging prominences at proximal end of tibia; upper surfaces are concave for articulation with femur
Intercondylar eminence	Upward projection on articular surface between condyles
Crest	Sharp ridge on anterior surface
Tibial tuberosity	Projection in midline on anterior surface
Popliteal line	Ridge that spirals downward and inward on posterior surface of upper third of tibial shaft
Medial malleolus	Rounded downward projection at distal medial end of tibia; forms prominence on inner surface of ankle
Fibula (Fig. 27-3)	Long, slender bone of lateral side of lower leg
Lateral malleolus	Rounded prominence at distal lateral end of fibula; forms prominence on outer surface of ankle
Tarsals (Fig. 27-3)	Seven bones that form the ankle of the foot (**calcaneus; talus; navicular; first, second,** and **third cuneiforms;** and **cuboid)**
Calcaneus	Heel bone
Talus	Uppermost of tarsals; articulates with tibia and fibula; boxed in by medial and lateral malleoli
Metatarsals (Fig. 27-3)	Long bones of the sole of the foot; numbered one through five from big toe to little toe
Phalanges (Fig. 27-3)	Miniature long bones of toes; two in each great toe (**hallux),** three in other toes (proximal, medial, and distal), numbered one through five from big toe to the little toe

FIG. 27-2 Tibia and fibula (anterior view).

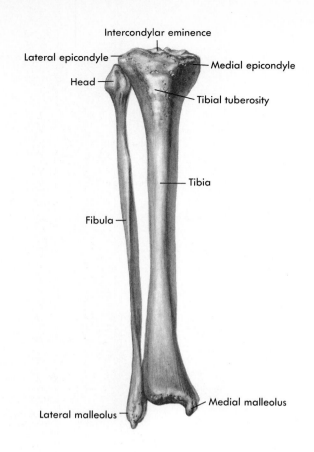

Intercondylar eminence

Lateral epicondyle

Head

Medial epicondyle

Tibial tuberosity

Tibia

Fibula

Medial malleolus

Lateral malleolus

FIG. 27-3 Tarsals, metatarsals, and phalanges. **A**, Dorsal view. **B**, Medial view.

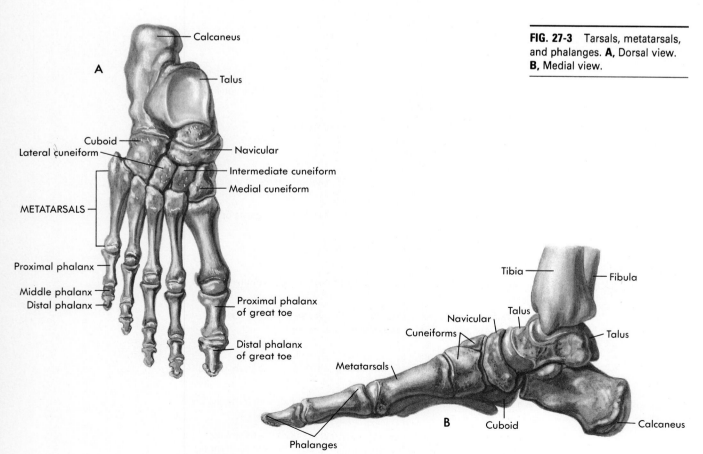

A

Calcaneus

Talus

Cuboid

Navicular

Lateral cuneiform

Intermediate cuneiform

Medial cuneiform

METATARSALS

Proximal phalanx

Middle phalanx

Distal phalanx

Proximal phalanx of great toe

Distal phalanx of great toe

Tibia

Fibula

Navicular

Talus

Cuneiforms

Talus

Metatarsals

Phalanges

Cuboid

Calcaneus

B

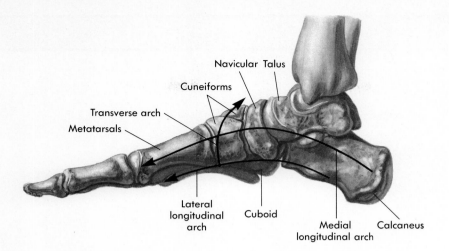

FIG. 27-4 Arches of foot.

To support the weight of the body and provide leverage during walking, the bones of the foot are arranged into arches (Fig. 27-4). The arches include the following:

Longitudinal arches	Tarsals and metatarsals so arranged as to form arch from front to back of foot
Medial arches	Formed by calcaneus, talus, navicular, cuneiforms, and three medial metatarsals
Lateral arches	Formed by calcaneus, cuboid, and two lateral metatarsals
Transverse arch	Metatarsals and distal row of tarsals so articulated as to form arch across foot; bones kept in two arched positions by means of powerful ligaments in sole of foot and by muscles and tendons

PROCEDURE A   LOWER EXTREMITIES

1. Use the individual bones and the model skeleton to identify and describe the bones and bone markings of the lower extremities.

PROCEDURE B   ARCHES OF THE FOOT

1. Use the skeleton to identify the bones associated with arches of the foot.

1. What is the function of the following appendicular skeletal markings?

   Greater trochanter

   Linea aspera

   Tibial tuberosity

2. What bone is the strongest of the skeletal system?

3. Which area of the femur is fairly commonly fractured in the elderly?

4. Where does development of the patella occur?

5. During walking, which tarsal bone initially bears the entire weight?

6. Explain which bones form the four foot arches.

*Use all references and materials at your disposal to answer these review questions.

X-RAY FRACTURE IDENTIFICATION AND FRACTURE REPAIR

28

OBJECTIVES

1 Define a fracture.
2 Describe the types of fractures.
3 Explain bone remodeling.

MATERIALS

x-ray films of fractures   x-ray films of remodeling

Callus
Colles' fracture
Comminuted
Complete
Compound
Fracture
Fracture hematoma
Greenstick
Impacted
Partial
Pott's fracture
Remodeling
Simple
Spiral
Transverse

A **fracture** is any break in a bone. The following terms may be used to classify fractures:

Partial (Fig. 28-1, *A*)	Incomplete break across bone
Complete (Fig. 28-1, *B*)	Complete break across bone
Simple (Fig. 28-1, *A*)	Fracture does not break skin
Compound (Fig. 28-1, *C*)	Fracture breaks skin
Comminuted (Fig. 28-1, *D*)	Bone is fragmented and splintered
Greenstick (Fig. 28-1, *E*)	Only occurs in children; one side of bone breaks and other side bends
Spiral (Fig. 28-1, *F*)	Bone is twisted apart
Transverse (Fig. 28-1, *D*)	Fracture at right angle to long bone
Impacted (Fig. 28-1, *G*)	One end of breakage is driven into other end
Pott's (Fig. 28-1, *H*)	Breakage of distal end of fibula and tibia
Colles' (Fig. 28-1, *I*)	Breakage of distal end of radius

When a fracture occurs, the ends of the blood vessels coagulate and form a **fracture hematoma.** A **callus** or new bone tissue will then form, linking the fractured areas. As new bone tissue forms, **remodeling** occurs. This is when dead portions of broken fragments are gradually reabsorbed.

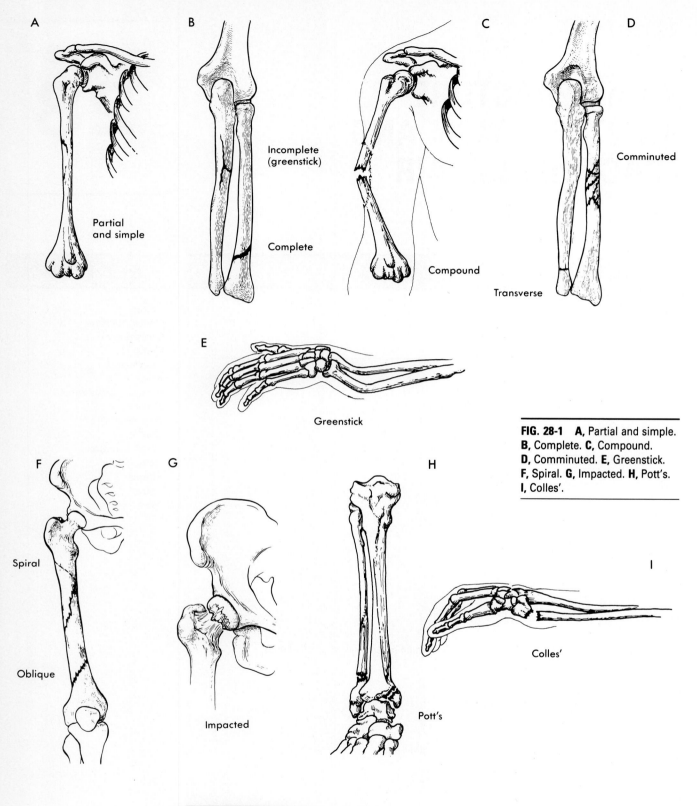

A

Partial
and simple

B

Incomplete
(greenstick)

Complete

C

Compound

D

Comminuted

Transverse

E

Greenstick

FIG. 28-1 **A,** Partial and simple.
B, Complete. **C,** Compound.
D, Comminuted. **E,** Greenstick.
F, Spiral. **G,** Impacted. **H,** Pott's.
I, Colles'.

F

Spiral

Oblique

G

Impacted

H

Pott's

I

Colles'

PROCEDURE A FRACTURES

1. Place various x-ray films over a light and describe the break.
2. Classify each fracture.

PROCEDURE B FRACTURE REPAIR

1. Place various x-ray films over a light and describe the stages of fracture repair.

1. Define the following terms:

   Fracture hematoma

   Callus

   Remodeling

2. How long does it take for a fracture hematoma to form?

3. Compare the external callus and internal callus.

4. Explain the process by which a fracture is repaired.

*Use all references and materials at your disposal to answer these review questions.

5. Place the correct fracture name below the diagram.

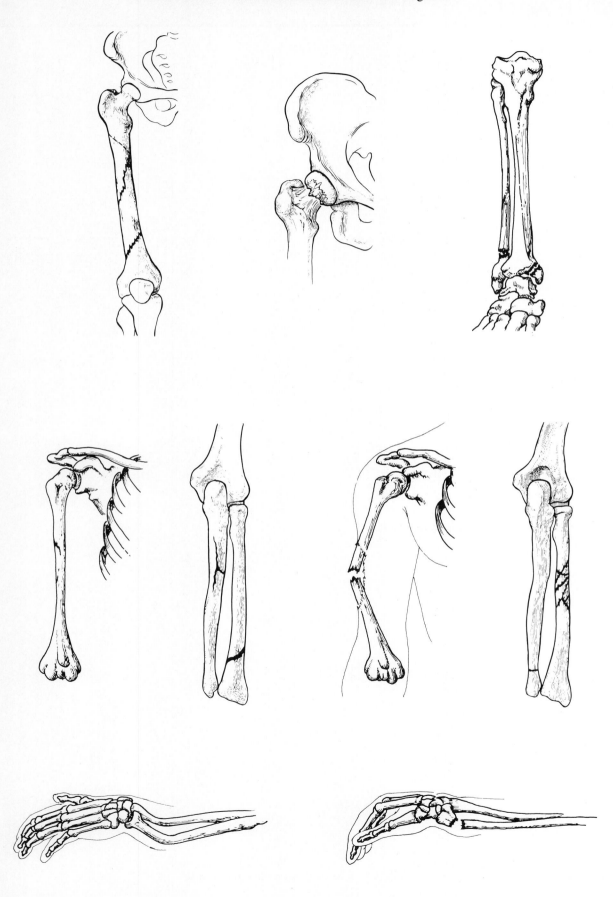

ARTICULATIONS

29

OBJECTIVES

1 Define an articulation.
2 Compare the structure, kind of movement, and location of fibrous, cartilaginous, and synovial joints.
3 Describe the movements of the various synovial joints.

MATERIALS

human skeleton

The terms **joint** and **articulation** are used to describe where bones meet. Articulations may be classified according to the degree of movement: **synarthroses,** or immovable joints; **amphiarthroses** or slightly movable joints; and **diarthroses,** or freely movable joints.

Further, joints may be classified according to the presence or absence of a space between the bones and the type of connective tissue binding the bones. **Fibrous joints** have no space between the bones, and they are bound together by **fibrous connective tissue. Cartilaginous joints** have no space between the bones and are bound together by **cartilage. Synovial joints** have a fluid filled space between the bones. Since the bones are not cemented together by tissues, free mobility is allowed.

Using both the anatomy and the movement of joints, articulations may be classified as follows:

IDENTIFICATION	DESCRIPTION	MOVEMENT	EXAMPLES
Fibrous			
Suture	Found only between bones of skull; articulating bones separated by thin layer of fibrous tissue	Synarthrotic	Lamboidal suture between occipital and parietal bones
Syndemosis (Fig. 29-1)	Articulating bones united by a considerable amount of fibrous tissue	Amphiarthrotic	Distal ends of tibia and fibula
Gomphosis	A thin layer of fibrous tissue is at the junction of a cone-shaped peg fitting into a socket; periodontal ligaments provide slight movement during mastication (chewing)	Synarthrotic	Roots of teeth in alveolar processes

KEY TERMS

Amphiarthroses
Articulation
Ball-and-socket
Biaxial
Cartilage
Cartilaginous joint
Diarthroses
Ellipsoidal
Fibrous connective tissue
Fibrous joint
Gliding
Gomphosis
Hinge
Joint
Monoaxial
Pivot
Saddle
Suture
Symphysis
Synarthroses
Synchondrosis
Syndesmosis
Synovial joint
Triaxial

FIG. 29-1 Syndesmosis.

FIG. 29-2 Synchondroses.

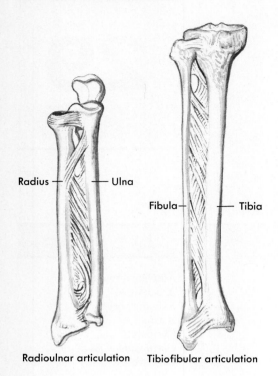

Radius — Ulna

Fibula — Tibia

Radioulnar articulation   Tibiofibular articulation

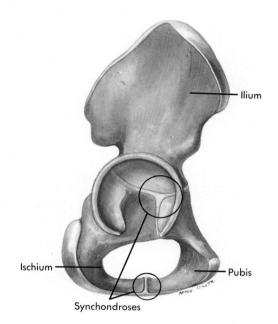

Ilium

Ischium

Pubis

Synchondroses

AFTER SCHAEFFER

IDENTIFICATION	DESCRIPTION	MOVEMENT	EXAMPLES
Cartilaginous			
Synchondrosis (Fig. 29-2)	Connecting material is hyaline cartilage	Synarthrotic	Temporary joint between diaphysis and epiphysis of long bone; fusion of ilium, ischium, and pubis in growing child
Symphysis	Connecting material is broad, flat disc of fibrocartilage	Amphiarthrotic	Intervertebral joints and symphasis pubis
Synovial (Fig. 29-3)	An articular capsule consisting of ligaments on the outside and a fluid-filled synovial membrane on the inside (between articulating surfaces) comprises this type of joint		
Gliding	Articulating surfaces, usually flat	Biaxial (permits movement in two planes)	Intercarpal and intertarsal joints; articular process between synovial joints

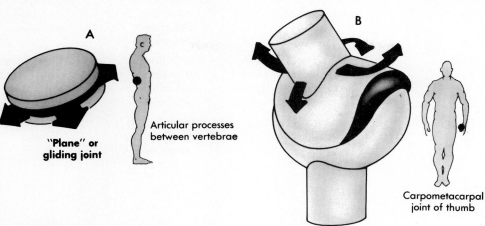

A

"Plane" or gliding joint

Articular processes between vertebrae

B

Saddle joint

Carpometacarpal joint of thumb

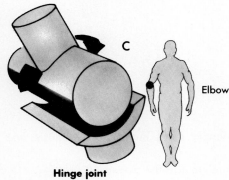

C

Hinge joint

Elbow

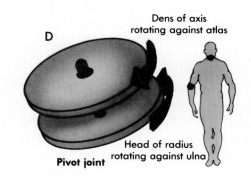

D

Dens of axis rotating against atlas

Head of radius rotating against ulna

Pivot joint

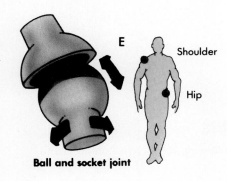

E

Shoulder

Hip

Ball and socket joint

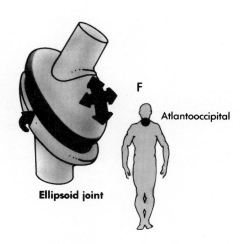

F

Atlantooccipital

Ellipsoid joint

IDENTIFICATION	DESCRIPTION	MOVEMENT	EXAMPLES
Hinge	Spool-like surface fits into a concave surface	Monoaxial (permits movement in one plane)	Elbow, knee, ankle, and interphalangeal joints
Pivot	Rounded, pointed, or concave surface fits into a ring formed partly by bone and partly by ligament	Monaxial (rotation)	Atlantoaxial and radioulnar joints

IDENTIFICATION	DESCRIPTION	MOVEMENT	EXAMPLES
Ellipsoidal	Oval-shaped condyle fits into elliptical cavity	Biaxial (flexion-extension, abduction-adduction)	Radiocarpal joint, atlantooccipital
Saddle	Articular surfaces concave in one direction and convex in opposite direction	Biaxial (flexion-extension, abduction-adduction)	Carpometacarpal joint of thumb
Ball-and-socket	Ball-like surface fits into a cuplike depression	Triaxial (permits movement in almost any direction)	Shoulder and hip joints

PROCEDURE A FIBROUS JOINTS

1. Locate and name the bones that form the following joints: sutures, syndesmosis, and gomphosis.
2. Describe the structure of the articulation and relate it to the movement.

PROCEDURE B CARTILAGINOUS JOINTS

1. Locate and name the bones that form the following joints: synchondrosis and symphysis.
2. Describe the structure of the articulation and relate it to the movement.

PROCEDURE C SYNOVIAL JOINTS

1. Locate and name the bones that form the following joints: gliding, hinge, pivot, ellipsoidal, saddle, and ball-and-socket.
2. Describe the structure of the articulation and relate it to the movement.

1. Define the following terms:

   Articulation

   Synarthroses

   Amphiarthroses

   Diarthroses

   Fibrous joint

   Cartilaginous joint

   Synovial joint

2. Give an example of the following articulations:

   Syndesmosis

   Gomphosis

   Suture

   Synchondrosis

   Symphysis

3. Give an example of the following movements:

   Gliding

   Hinge

   Pivot

   Ellipsoidal

   Saddle

   Ball-and-socket

*Use all references and materials at your disposal to answer these review questions.

30 SYNOVIAL JOINTS

KEY TERMS

Articular capsule
Bursae
Bursitis
Coxal joint
Dense connective tissue
Fibrous capsule
Humeroscapular joint
Ligament
Loose connective tissue
Periosteum
Synovial joint
Synovial membrane
Tibiofemoral joint

OBJECTIVES

1 Describe the structural features of an intact diarthritic joint.
2 List the structural components of a typical diarthrotic joint.
3 Discuss the appearance and physical characteristics of each anatomical component of a synovial joint.

MATERIALS

intact beef or lamb knee,
   shoulder, and hip joints
beef or lamb knee, shoulder,
   and hip joints sawed in half
   longitudinally

dissecting equipment

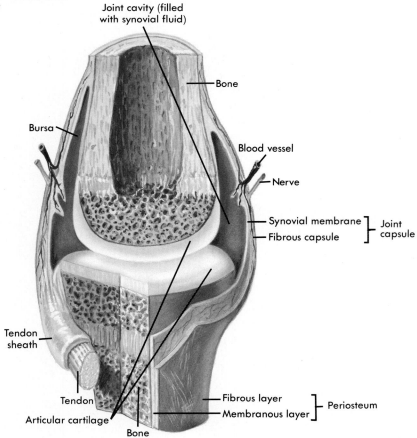

FIG. 30-1   Structure of synovial joint.

110

When articulating bones form a cavity, the articulation is called a **synovial joint** (Fig. 30-1). The articulation is enclosed by the **articular (joint) capsule,** which is composed of a **dense connective tissue outer layer** called the **fibrous capsule** and a **loose connective tissue inner layer** called the **synovial membrane.**

The fibrous capsule prevents dislocation of the joint as well as flexibility for movement. Fibrous capsules arranged in parallel bundles form the **ligaments,** which attach to the **periosteum** of the articulating bones. The major synovial joints include the tibiofemoral joint (Fig. 30-2), humeroscapular joint, and coxal joint (Fig. 30-3).

The synovial membrane lubricates the joint and provides nourishment for the articulating cartilage. Similar saclike membrane structures called **bursae** are located between bones and other tissue in areas of high stress to prevent friction. Inflammation of the bursae is called **bursitis.**

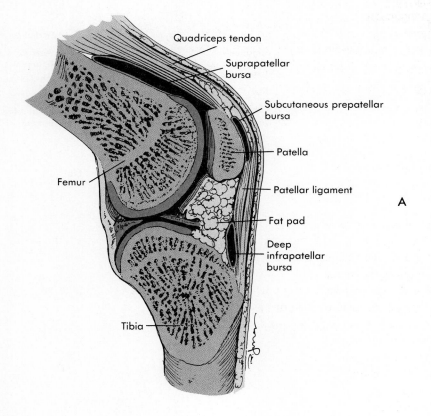

FIG. 30-2 Tibiofemoral joint.
A, Sagittal section.

Quadriceps tendon

Suprapatellar bursa

Subcutaneous prepatellar bursa

Patella

Femur

Patellar ligament

Fat pad

Deep infrapatellar bursa

A

Tibia

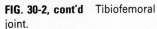

FIG. 30-2, cont'd Tibiofemoral joint.
B, Anterior superficial view.
C, Anterior deep view.
D, Posterior superficial view.
E, Posterior deep view.

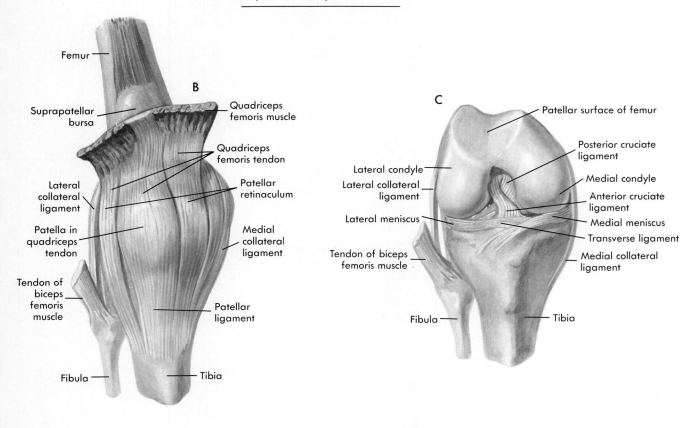

B

Femur

Suprapatellar bursa

Lateral collateral ligament

Patella in quadriceps tendon

Tendon of biceps femoris muscle

Fibula

Tibia

Quadriceps femoris muscle

Quadriceps femoris tendon

Patellar retinaculum

Medial collateral ligament

Patellar ligament

C

Patellar surface of femur

Posterior cruciate ligament

Medial condyle

Anterior cruciate ligament

Medial meniscus

Transverse ligament

Medial collateral ligament

Lateral condyle

Lateral collateral ligament

Lateral meniscus

Tendon of biceps femoris muscle

Fibula

Tibia

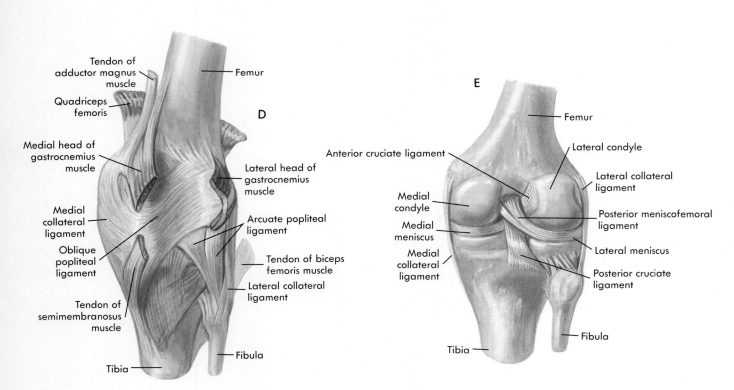

D

Tendon of adductor magnus muscle

Quadriceps femoris

Medial head of gastrocnemius muscle

Medial collateral ligament

Oblique popliteal ligament

Tendon of semimembranosus muscle

Tibia

Femur

Lateral head of gastrocnemius muscle

Arcuate popliteal ligament

Tendon of biceps femoris muscle

Lateral collateral ligament

Fibula

E

Anterior cruciate ligament

Medial condyle

Medial meniscus

Medial collateral ligament

Femur

Lateral condyle

Lateral collateral ligament

Posterior meniscofemoral ligament

Lateral meniscus

Posterior cruciate ligament

Fibula

Tibia

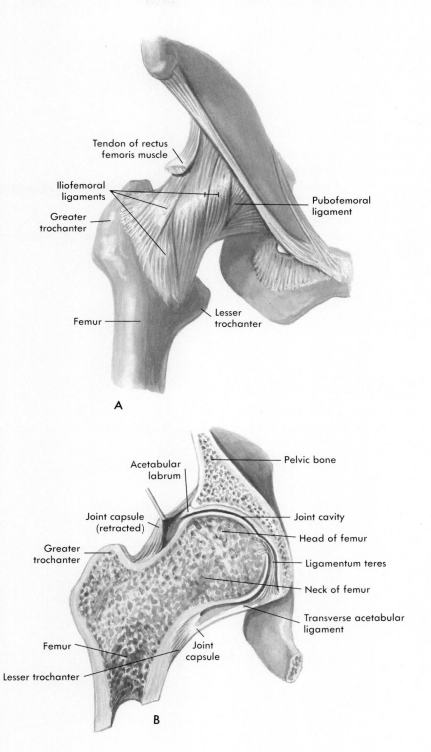

FIG. 30-3 Coxal joint.
A, Anterior view. B, Frontal section.

Tendon of rectus
femoris muscle

Iliofemoral
ligaments

Greater
trochanter

Pubofemoral
ligament

Femur

Lesser
trochanter

A

Acetabular
labrum

Joint capsule
(retracted)

Greater
trochanter

Pelvic bone

Joint cavity

Head of femur

Ligamentum teres

Neck of femur

Transverse acetabular
ligament

Femur

Joint
capsule

Lesser trochanter

B

PROCEDURE A   TIBIOFEMORAL (KNEE) JOINT

1. Feel the intact joint and make observations.
2. Using the longitudinal section of the joint, feel the membrane and note observations.
3. Using the intact knee joint, remove any muscle tissue to identify the following ligaments:
   Patellar
   Lateral patellar
   Medial patellar
   Fibular collateral
   Tibial collateral
   Oblique popliteal
   Arcuate popliteal
   Anterior ligament of head of fibula
   Posterior ligament of head of fibula
4. Remove the periosteum from the lower bone and pull it toward the articulating bone. Record observations.
5. Cut open the patellar ligament and locate the patella.
6. Cut away the outer ligaments to observe the anterior and posterior cruciate and transverse ligaments.

PROCEDURE B   COXAL (HIP) JOINT

1. Feel the intact joint and note observations.
2. Remove any muscle tissue to identify the following ligaments:
   Iliofemoral
   Pubofemoral
   Ischiofemoral

1. Draw and label a typical synovial joint.

2. Describe the function of the following joint and associated structures:

   Articular cartilage

   Articular capsule

   Fibrous capsule

   Synovial membrane

   Menisci

   Bursae

   Ligament

3. Define the following homeostatic imbalances:

   Rheumatoid arthritis

   Bursitis

   Tendinitis

   Gouty arthritis

   Torn cartilage

*Use all references and materials at your disposal to answer these review questions.

4. Use the following diagrams to place the correct ligament name next to the appropriate label line.

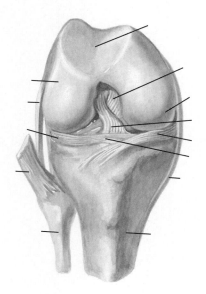

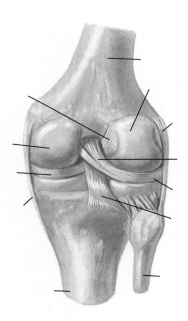

5. What limits movement at synovial joints?

DIARTHROTIC JOINT MOVEMENT

31

OBJECTIVES

1 Compare the movements at various synovial joints.
2 Compare the structure of diarthrotic joints with the function.

KEY TERMS

Abduction
Adduction
Anatomical position
Angular movement
Circumduction
Depression
Dorsiflexion
Elevation
Eversion
Extension
Flexion
Hyperextension
Inversion
Lateral rotation
Medial rotation
Plantar flexion
Pronation
Protraction
Retraction
Rotation
Special movements
Supination

MATERIALS

laboratory partner

Movements that occur at synovial joints are related to the structure of the joint. Movements of diarthrotic joints are always relative to the anatomical position (standing with palms of the hands and the feet facing forward) (Fig. 31-1).

Gliding movement provides simple back-and-forth as well as side-to-side movement over two surfaces. Therefore, these are biaxial.

An increase or decrease in the angle between bones is called **angular movement** (Fig. 31-2). **Flexion** decreases the angle between articulating bones. Flexion of the foot (pointing the toes toward the skin) is called **dorsiflexion. Extension** increases the angle between articulating bones. Overextension of the anatomical position is called **hyperextension.** Extension of the foot (standing on the toe or ball of the foot) is called **plantar extension. Abduction** is the movement of a bone away from the body's midline. **Adduction** is the movement of a bone toward the body's midline.

When no other motion is occurring, the movement around an axis is called **rotation** (Fig. 31-3). **Medial rotation** is the movement of the anterior (ventral) surface of the bone and rotates toward the midline, whereas **lateral rotation** is the movement of the anterior surface of the bone and rotates away from the midline. Since rotation only occurs in one plane, it is monoaxial.

Movement of the distal end of a bone in a circular pattern while the proximal end of the bone is stationary is called **circumduction** (Fig. 31-4). This is triaxial.

Besides the above mentioned movements, there are many **special movements** (Fig. 31-5). **Inversion,** is the medial turning in of the sole of the foot. **Eversion** is lateral movement of the sole to anatomical position. Movement on a plane parallel to the ground by the clavicle or mandible is called **protraction.** The return position involves **retraction. Supination** is the rotation of the palm so that it faces anteriorly, and **pronation** is rotation of the palm so that it faces posteriorly. **Elevation** is upward movement of body parts, and the return to anatomical position is called **depression.**

PROCEDURE A   DIARTHROTIC JOINT MOVEMENTS

1. Using your laboratory partner, demonstrate all diarthrotic joint movements.

FIG. 31-1 Anatomical position.

FIG. 31-2 **A,** Abduction and adduction. **B,** Flexion and extension.

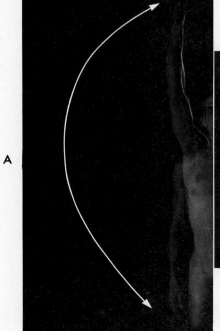

A

B

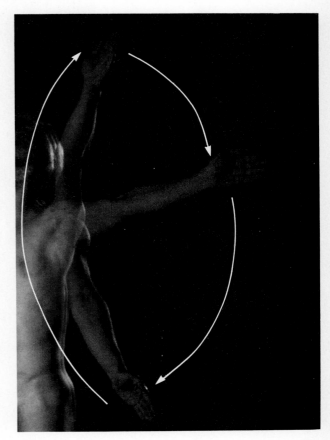

FIG. 31-2,cont'd   C, Dorsiflexion and plantar flexion.

C

FIG. 31-3   Rotation.

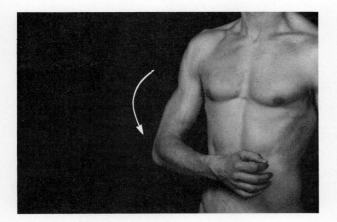

FIG. 31-4   Circumduction.

FIG. 31-5 **A,** Eversion and inversion. **B,** Protraction and retraction. **C,** Pronation and supination. **D,** Elevation and depression.

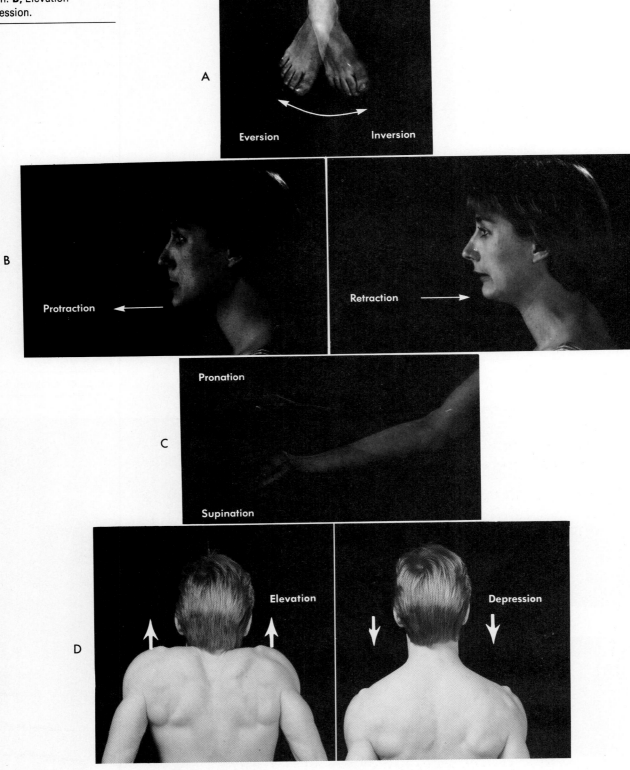

A

Eversion   Inversion

B

Protraction ←   Retraction →

Pronation

C

Supination

D

Elevation ↑↑   Depression ↓↓

1. Define the following diarthrotic joint movements:

   Flexion

   Extension

   Dorsiflexion

   Plantar flexion

   Adduction

   Abduction

   Eversion

   Inversion

   Supination

   Pronation

2. Label the following diagrams with correct movement.

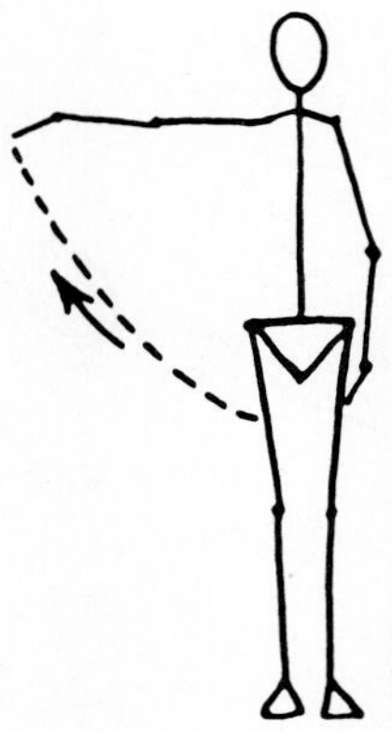

Continued.

*Use all references and materials at your disposal to answer these review questions.

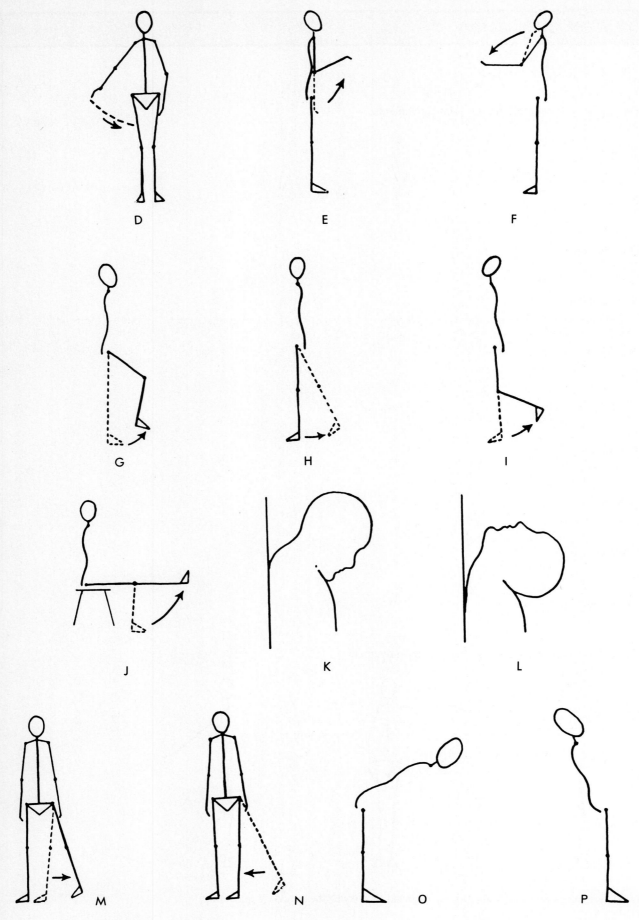

D

E

F

G

H

I

J

K

L

M

N

O

P

122

INTEGUMENTARY SYSTEM
Histology and physiology

32

OBJECTIVES	KEY TERMS

OBJECTIVES

1 Compare the epidermis and dermis.
2 List the layers of the epidermis and describe their functions.
3 Name the cutaneous receptors and explain the functions of each.

MATERIALS

cutaneous prepared slide   toothpick
compound microscope   marker
horsehair   ruler
cold cylinder

KEY TERMS

Cutaneous receptors
Cutis
Dermis
Elidin
Epidermis
Free nerve endings
Hair end organs
Integumentary system
Keratin
Keratohyalin
Krause's end bulb
Meissner corpuscles
Merkel's disks
Pacini corpuscles
Ruffini's corpuscles
Stratum basale
Stratum corneum
Stratum granulosum
Stratum lucidum
Stratum spinosum

The **integumentary system** includes the **skin,** or **cutis,** and the associated structures such as the **hair, nails, glands,** and some **specialized receptors.** The skin has two basic parts, the **epidermis** and the **dermis.** Each region may be described as follows:

Epidermis (Fig. 32-1 and Color Plate 9)	Outer region of skin; stratified squamous epithelium with four to five layers, depending on location
Stratum basale	Bottom layer of epidermis; columnar cells that undergo continuous mitosis and move upward, changing shape; as cells move upward, nuclei degenerate, and cells die and shed from top layer of epidermis
Stratum spinosum	Several rows of living polyhedral cells
Stratum granulosum	Several rows of squamal cells containing **keratohyalin;** this compound forms **elidin,** a translucent substance eventually transformed to **keratin,** a waterproofing protein
Stratum lucidum	Several rows of squamal cells that are dead and appear clear because of the substance **elidin,** derived from keratohyalin
Stratum corneum	Many rows of squamal cells that are dead; elidin has transformed into **keratin**
Dermis (Fig. 32-1 and Color Plate 9)	Lower region of skin composed of connective tissue with collagenous and elastic fibers; embedded in this layer are blood vessels, nerves, glands, and hair follicles

FIG. 32-1 Epidermal strata.

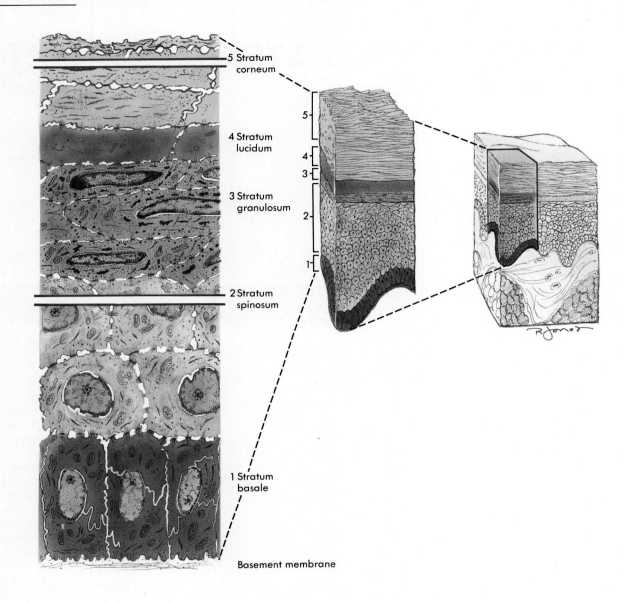

5 Stratum corneum

4 Stratum lucidum

3 Stratum granulosum

2 Stratum spinosum

1 Stratum basale

Basement membrane

The cutaneous receptors respond to **tactility** (touch), **pressure, temperature,** and **pain stimuli.** Though cutaneous receptors are somewhat specific, density along the body surface is not uniform. **Cutaneous receptors** include the following (Fig. 32-2):

Free nerve endings	Throughout the skin, sensitive to touch, pressure, and pain
Meissner's corpuscles	Found in fingertips, lips, eyelids, and external genitalia; encapsulated nerve endings; sensitive to vibrations, touch, and pressure
Merkel's disks	Found in epidermis; modified Meissner's corpuscles; sensitive to touch
Hair end organs	Found attached to hair follicles; sensitive to touch or movement of hair follicles
Ruffini's corpuscles	Found in connective tissue; sensitive to touch and heat
Krause's end bulb	Found in mucocutaneous areas (lips and external genitalia) and in relation to hairs
Pacini corpuscles	Found in subcutaneous tissue of palms, soles, digits, periosteum, mesentery, tendons, ligaments, and external genitalia

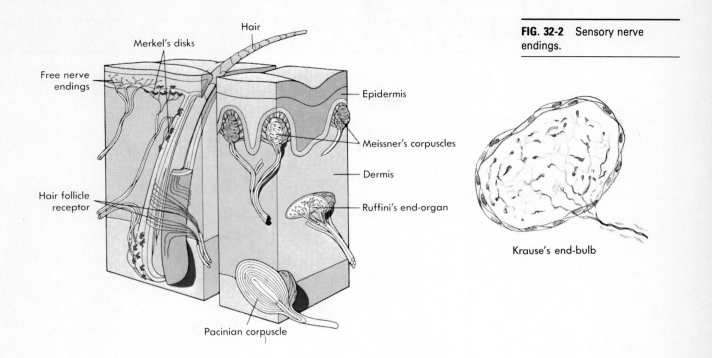

FIG. 32-2 Sensory nerve endings.

Krause's end-bulb

PROCEDURE A   INTEGUMENTARY HISTOLOGY

1. Using high power, focus on a prepared cutaneous slide.
2. Identify the epidermis layers and describe their functions.
3. Carefully examine the dermis, noting blood vessels, glands, nerves, and hair follicles.
4. Draw and label the epidermis and dermis.

PROCEDURE B   CUTANEOUS SENSATIONS AND DENSITY

1. Have your laboratory partner close his or her eyes. Gently pass a horsehair across the back of the forearm to recognize the touch stimuli.
2. Repeat the procedure several times until the stimulus is easily recognized.
3. Repeat the procedure using a cold cylinder to recognize temperature stimuli and a toothpick to recognize pain stimuli.
4. Using a felt-tip pen, draw a 1-inch square on the back forearm. Then divide the square into 16 smaller squares.
5. Have your partner close his or her eyes while you apply each stimulus to each square. Record the number of times each stimulus is felt.

1. Draw and label the layers of the epidermis.

2. Describe the structures of the dermis.

3. Name and explain the function of the cutaneous receptors.

4. Explain the following disorders of the integumentary system:

   Acne

   Impetigo

   Systemic lupus erythematosus

   Psoriasis

   Sunburn

5. Describe first-degree, second-degree, and third-degree burns.

*Use all references and materials at your disposal to answer these review questions.

INTEGUMENTARY SYSTEM
Removal of animal skin

33

OBJECTIVES

1 Removal of dissecting animal skin for muscular identification.

MATERIALS

dissecting animal   dissecting equipment

Before muscular examination, removal of the skin is necessary (Fig. 33-1).

PROCEDURE A   REMOVAL OF SKIN FROM DISSECTING ANIMAL

1. Place the dissecting animal's dorsal surface on a dissecting tray.
2. Make an incision along the ventral midline from the opening inferior to the chin (previously made for latex injection) to the lower abdominal region.
3. Make lateral cuts: from the inferior surface of the chin to the back of the ears; from the clavicle to the distal region of the forelimb; along the thoracic region; along the abdominal region; along the caudal region; along the hindlimb to the distal region.
4. Carefully separate the skin from the muscles using a blunt probe. The thin transparent membrane is the superficial fascia.
5. Once the skin from the ventral side has been removed, place the dissecting animal's ventral side on the dissecting tray. Remove the skin from the dorsal side.
6. Retain the skin so that it may be wrapped around the animal between investigations. This will prevent drying out.
7. Carefully remove the fascia so that each muscle may be distinguished.

FIG. 33-1 **A,** Ventral view. **B,** Dorsal view.

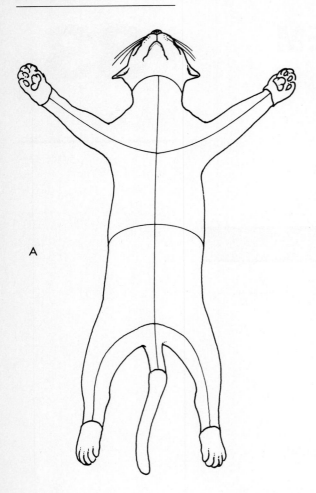

A

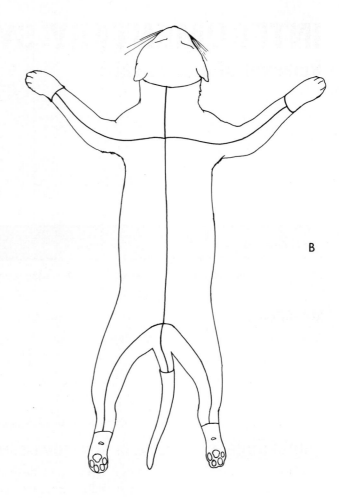

B

MUSCULAR SYSTEM
Histology

34

OBJECTIVES

1 Describe the relationship between deep fascia, epimysium, perimysium, and endomysium.
2 Characterize a typical skeletal muscle fiber.
3 Identify the microscopic components of skeletal muscle tissue.

KEY TERMS

Actin
Deep fascia
Endomysium
Epimysium
Fasciculi
Muscle fibers
Myofibril
Myofilament
Myosin
Perimysium
Sarcolemma
Sarcomere
Skeletal muscles
Striated muscle tissue
Tendon

MATERIALS

dissecting animal
dissecting equipment
skeletal muscle tissue slide

compound microscope
dissecting microscope

Skeletal muscles attach to the skeleton. As muscle contraction occurs, the body may be put in motion or held in a standing position.

Around each skeletal muscle is a layer of fibrous dense connective tissue called **deep fascia** (Fig. 34-1). It is the deep fascia that provides protection, holds muscles together, and separates them. An extension of the deep fascia, which further protects the muscles, is called the **epimysium.** The epimysium "compartmentalizes" each muscle. An extension of the epimysium into the muscle is called the **perimysium.** The perimysium protects and bundles groups of *muscle fibers (cells)* called **fasciculi.** Within the fasciculi each muscle fiber is protected by an extension of the perimysium called the **endomysium.** It is the extension of all of these connective tissues beyond the muscle cells that make up a **tendon,** which attaches the muscle to the periosteum of the bone.

A single muscle fiber (Color Plate 10 and Fig. 34-2) is characterized as having many nuclei lying close to the **sarcolemma,** or plasma membrane. Each muscle fiber contains bundles of protein myofilaments called **myofibrils.** Thin myofilaments are called **actin,** and thick myofilaments are called **myosin.** The myofibrils are arranged in short groups called **sarcomeres.** Microscopically, sarcomeres appear striated, thereby giving skeletal muscle the name **striated muscle tissue.**

During muscle contraction, the myofilaments of myofibrils slide across one another. Because skeletal muscles may be contracted voluntarily, skeletal muscle tissue has yet another name, **voluntary muscle tissue.**

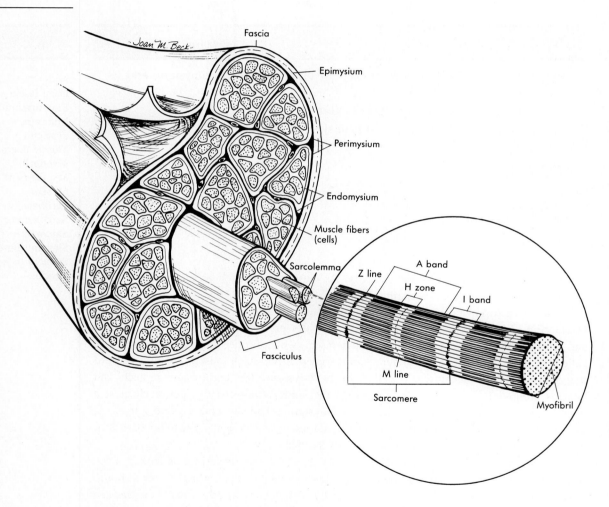

FIG. 34-1 Relationship between muscle fibers and associated structures.

Fascia

Joan M Beck

Epimysium

Perimysium

Endomysium

Muscle fibers (cells)

Sarcolemma

Fasciculus

Z line

A band

H zone

I band

M line

Sarcomere

Myofibril

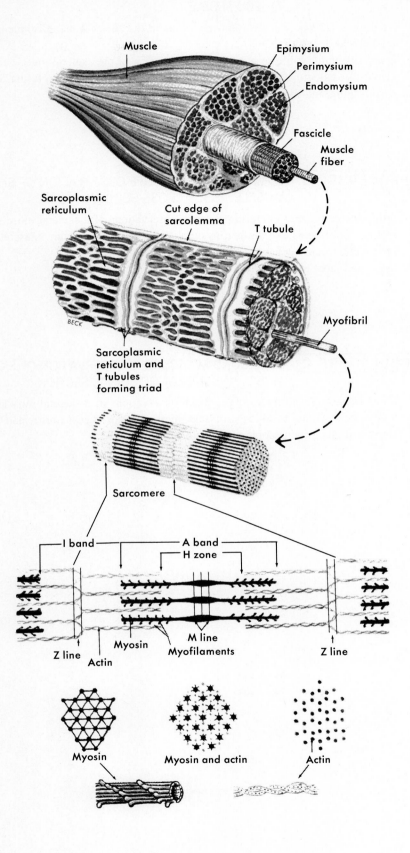

FIG. 34-2 Structure of skeletal muscle.

Muscle

Epimysium

Perimysium

Endomysium

Fascicle

Muscle fiber

Sarcoplasmic reticulum

Cut edge of sarcolemma

T tubule

BECK

Myofibril

Sarcoplasmic reticulum and T tubules forming triad

Sarcomere

I band

A band

H zone

Myosin

M line

Myofilaments

Z line

Actin

Z line

Myosin

Myosin and actin

Actin

PROCEDURE A DISSECTING MICROSCOPE OBSERVATION OF SKELETAL MUSCLE TISSUE

1. Remove a section of skeletal muscle tissue at the site of bone attachment. Observe the tendon and deep fascia.
2. Use a dissecting knife to remove a cross-section of the skeletal muscle tissue.
3. Using the dissecting microscope, make observations of the deep fascia, epimysium, perimysium, and fasciculi.

PROCEDURE B COMPOUND MICROSCOPE OBSERVATION OF SKELETAL MUSCLE TISSUE CROSS-SECTION

1. Using the compound microscope, make observations of a skeletal muscle tissue cross-section slide to observe the perimysium, endomysium, fasciculi, muscle fiber sarcolemma, and nucleus.
2. Draw and label microscopic observations.

PROCEDURE C COMPOUND MICROSCOPE OBSERVATION OF SKELETAL MUSCLE TISSUE LONGITUDINAL SECTION

1. Using the compound microscope, make observations of a skeletal muscle tissue longitudinal section to observe muscle fibers, nucleus, sarcolemma, and striations.
2. Draw and label microscopic observations.

1. Place the correct term on the appropriate label line.

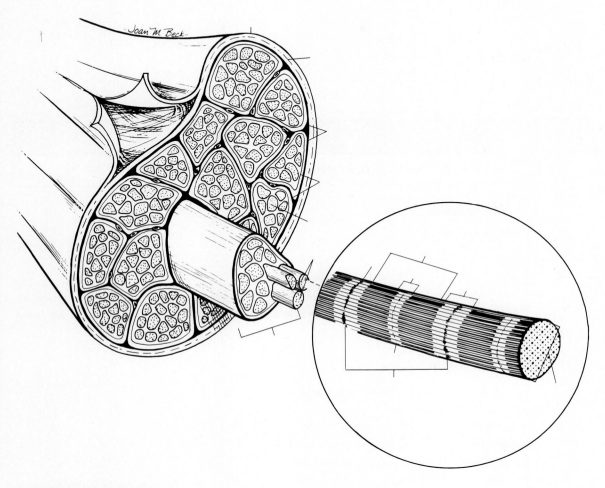

2. Define the following terms:

   Skeletal muscle
   Deep fascia
   Epimysium
   Perimysium
   Fasciculi
   Endomysium
   Tendon

3. Explain the sliding filament theory.

4. Why are the terms *striated muscle tissue* and *voluntary muscle tissue* applied to the skeletal muscle tissue?

*Use all references and materials at your disposal to answer these review questions.

35

MUSCULAR SYSTEM
Muscle action

KEY TERMS

Abductor
Adductor
Depressor
Dorsiflexor
Evertor
Extensor
Flexor
Invertor
Levator
Plantar flexor
Pronator
Rotator
Sphincter
Supinator
Tensor

OBJECTIVES

1 Identify various muscles according to their action.

MATERIALS

laboratory partner

Muscles may be named according to the direction, location, size, origin, insertion, or action of the muscle fibers. Muscles that are classified according to their action are listed below.

Muscle Classification	Action
Flexor (Fig. 31-2)	Decreases angle of joint
Extensor	Increases angle of joint
Abductor	Moves bone away from midline
Adductor	Moves bone toward midline
Levator (Fig. 31-5)	Produces upward movement
Depressor	Produces downward movement
Supinator	Turns palm upward or anteriorly
Pronator	Turns palm downward or posteriorly
Dorsiflexor (Fig. 31-2)	Flexes ankle joint and points toes superiorly
Plantar flexor	Extends ankle joint and points toes inferiorly
Invertor (Fig. 31-5)	Turns sole of foot inward
Evertor	Turns sole of foot outward
Sphincter (Fig. 31-6)	Decreases size of an opening
Tensor (Fig. 40-2)	Makes body part more rigid
Rotator (Fig. 31-3)	Moves bone around longitudinal axis

PROCEDURE A ACTION OF MUSCLES

1. Locate the following muscles and demonstrate how their name applies to their action.
   Flexor carpi radialis
   Extensor carpi ulnaris
   Abductor hallucis longus
   Adductor longus
   Levator scapulae
   Depressor labii inferioris
   Supinator
   Pronator teres
   Tibialis anterior
   Plantaris
   Tibialis anterior
   Peroneus tertius
   Orbicularis oculi
   Tensor fasciae latae
   Obturator

1. Explain the action of each of the following muscles:

   Flexor carpi radialis

   Extensor carpi ulnaris

   Abductor hallucis longus

   Adductor longus

   Levator scapulae

   Depressor labii inferioris

   Supinator

   Pronator teres

   Tibialis anterior

   Peroneus tertius

   Tensor fasciae latae

   Obturator

*Use all references and materials at your disposal to answer these review questions.

MUSCULAR SYSTEM
Facial expressions

36

OBJECTIVES

1 Describe the muscles involved with facial expressions.
2 Name the origin and insertion of the facial muscles.

MATERIALS

human muscle model   laboratory partner

Facial expressions occur because of the attachment of the skeletal muscles to the skin. Muscles of facial expressions are included in Table 36-1.

KEY TERMS

Buccinator
Corrugator supercili
Inferior labial depressor
Mentalis
Nasalis
Occipitofrontalis
Orbicularis oculi
Orbicularis oris
Platysma
Procerus
Risorius
Zygomaticus major
Zygomaticus minor

TABLE 36-1 FACIAL MUSCLES

MUSCLE	FUNCTION	ORIGIN	INSERTION
Occipitofrontalis	Raises eyebrows; moves scalp	Occipital bone	Tissue of eyebrow
Corrugator supercilii	Draws eyebrows together	Frontal bone (superciliary ridge)	Skin of eyebrow
Orbicularis oculi	Closes eye	Maxilla and frontal bones	Encircles eye and inserts at origin
Procerus	Forms horizontal wrinkle above nose	Bridge of nose	Skin over frontalis
Depressor labii inferiorus	Depresses lower lip	Mandible	Skin of lower lip and obicularis oris
Mentalis	Elevates and protrudes lower lip	Incisive fossa of mandible	Skin of chin
Nasalis	Dilates nostril	Maxilla	Ala of nose
Zygomaticus major	Abducts and elevates lip	Zygomatic bone	Angle of mouth
Zygomaticus minor	Abducts and elevates lip	Zygomatic bone	Upper lip
Orbicularis oris	Closes lips	Nasal septum, maxilla, and mandible	Fascia of lips
Platysma	Depresses lower lip; depresses mandible	Fascia of upper part of deltoid and pectoralis major	Mandible (lower border)
Buccinator	Flattens cheeks; compresses cheeks	Maxilla and mandible	Skin on sides of mouth
Risorius	Draws angle of mouth laterally	Fascia of platysma and masseter	Skin on sides of mouth and orbicularis oris

FIG. 36.1 Muscles of facial
expressions. **A.** Lateral view.
B. Anterior view.

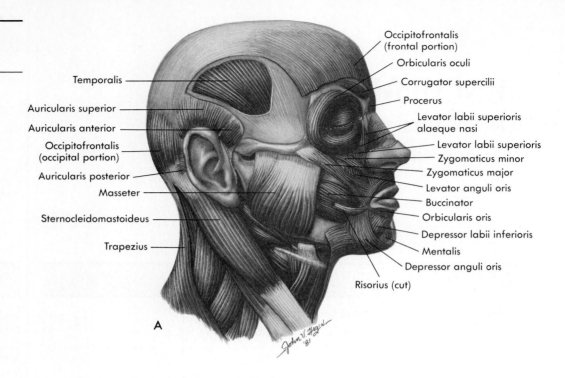

Temporalis

Auricularis superior

Auricularis anterior

Occipitofrontalis
(occipital portion)

Auricularis posterior

Masseter

Sternocleidomastoideus

Trapezius

Occipitofrontalis
(frontal portion)

Orbicularis oculi

Corrugator supercilii

Procerus

Levator labii superioris
alaeque nasi

Levator labii superioris

Zygomaticus minor

Zygomaticus major

Levator anguli oris

Buccinator

Orbicularis oris

Depressor labii inferioris

Mentalis

Depressor anguli oris

Risorius (cut)

A

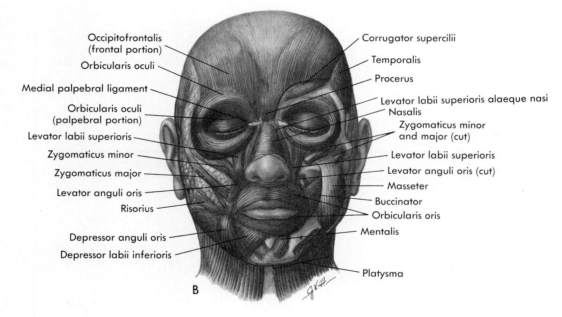

Occipitofrontalis
(frontal portion)

Orbicularis oculi

Medial palpebral ligament

Orbicularis oculi
(palpebral portion)

Levator labii superioris

Zygomaticus minor

Zygomaticus major

Levator anguli oris

Risorius

Depressor anguli oris

Depressor labii inferioris

Corrugator supercilii

Temporalis

Procerus

Levator labii superioris alaeque nasi

Nasalis

Zygomaticus minor
and major (cut)

Levator labii superioris

Levator anguli oris (cut)

Masseter

Buccinator

Orbicularis oris

Mentalis

Platysma

B

PROCEDURE A MUSCLES OF FACIAL EXPRESSIONS

1. Using the human model, locate each of the facial muscles and describe their function.
2. Name the origin and insertion of each facial muscle.
3. Using your laboratory partner, demonstrate the expressions made by the various facial
   muscles.

1. Which muscles are used to do the following?

   Smile

   Pucker the lips

   Blow

   Squint

   Frown

   Show surprise

   Pout

   Raise the eyebrows

2. Name the function, origin, and insertion of each of the following muscles:

   Occipitofrontalis

   Orbicularis oculi

   Depressor labii inferiorus

   Mentalis

   Zygomaticus major

   Obicularis oris

   Buccinator

*Use all references and materials at your disposal to answer these review questions.

37 MUSCULAR SYSTEM
Head, neck, thorax, and abdominal regions (ventral view)

OBJECTIVES

1 Identify the muscles of the head, neck, ventral thorax, and abdomen.
2 Describe the origin and insertion of the muscles of the head, neck, ventral thorax, and abdomen.
3 Explain the action of the muscles of the head, neck, ventral thorax, and abdomen.

MATERIALS

dissecting animal   dissecting equipment

Most muscles join articulating bones. The muscle tendon (a cord of connective tissue) attached to the stationary bone during contraction is called the **origin,** and the muscle tendon attached to the movable bone is called the **insertion.** Muscles of the head, neck, thorax, and abdominal regions are listed in Table 37-1 (Fig. 37-1). It is important to note that some muscles overlap other muscles.

PROCEDURE A   MUSCLES OF THE HEAD AND NECK

1. Identify the head and neck muscles.
2. Name the insertion and origin of each muscle.
3. Explain the action of each muscle.

PROCEDURE B   MUSCLES OF THE THORAX

1. Identify the superficial muscles of the ventral thorax.
2. Name the insertion and origin of each muscle.
3. Explain the action of each muscle.
4. *Transect* (make a right-angle cut to the muscle belly) the pectoantebrachialis, pectoralis major, pectoralis minor, and xiphihumeralis muscles to identify the scalenus, transversussternum costeria, serratus ventralis, subscapularis, teres major, and external intercostal muscles.
5. Identify, name the insertion and origin, and explain the function of each deep muscle of the ventral thorax.

PROCEDURE C   MUSCLES OF THE ABDOMEN

1. Identify the superficial muscles of the abdomen.
2. Name the insertion and origin of each muscle.
3. Explain the action of each muscle.
4. Transect the external oblique muscle to identify the internal oblique and transversus abdominis muscles.

FIG. 37-1 Head muscles.
A, Ventral view. **B,** Lateral view.

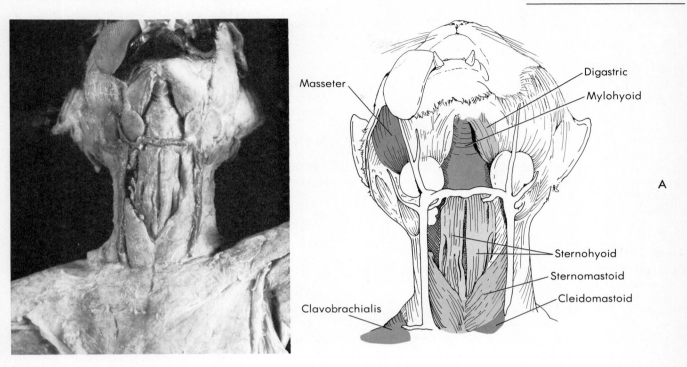

Masseter

Digastric

Mylohyoid

Sternohyoid

Sternomastoid

Cleidomastoid

Clavobrachialis

A

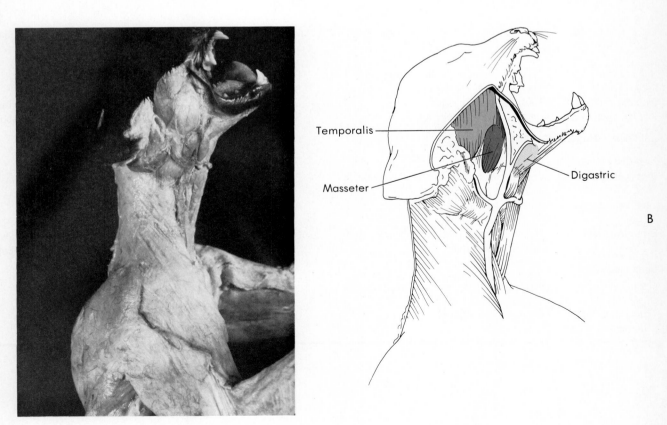

Temporalis

Masseter

Digastric

B

TABLE 37-1 MUSCLES OF THE HEAD, NECK, THORAX, AND ABDOMINAL REGIONS

MUSCLE	FUNCTION	ORIGIN	INSERTION
NECK AND HEAD MUSCLES (FIG. 37-1)			
Superficial muscles			
Masseter	Elevates and protracts mandible	Zygomatic	Mandible
Temporalis	Elevates and retracts mandible	Parietal temporal	Mandible
Digastric	Depresses and retracts mandible; elevates hyoid	Occipital temporal	Mandible
Mylohyoid	Depresses mandible; elevates tongue and floor of mouth	Body of mandible	Hyoid
Sternohyoid	Depresses hyoid	Sternum	Hyoid
Sternocleidomastoid and cleidomastoid	Flexes neck	Manubrium of sternum	Mastoid process of temporal bone
THORAX MUSCLES (FIG. 37-2)			
Superficial muscles			
Pectoantebrachialis (not found in humans)	Pulls humerus toward chest (adduction) and caudally	Anterior portion of sternum	Fascia of ulna
Pectoralis major	Pulls humerus toward chest (adduction) and caudally	Anterior portion of sternum; clavicle	Humerus
Pectoralis minor	Pulls humerus toward chest (adduction) and caudally	Mid and posterior sternum	Humerus
Xiphihumeralis	Pulls humerus toward chest (adduction) and caudally	Xiphoid process	Humerus
Deep muscles			
Scalenus (three separate muscles)	Elevates ribs	Transverse process of cervical vertebrae	Ribs
Transversuscostarum	Elevates ribs	Sternum	First rib
Serratus ventralis	Elevates ribs; rotates and protracts scapula	Lateral surface of upper right ribs; transverse process of last five cervical vertebrae	Dorsal margin of scapula
Subscapularis	Extends and medially rotates arm	Subscapularis	Humerus
Teres major	Rotates and flexes humerus	Scapula	Humerus
External intercostals	Elevates ribs; expiration	Inferior margin of rib	Superior border of rib below
Internal intercostals	Depresses ribs; inspiration	Superior margin of rib	Inferior border of rib above
ABDOMINAL MUSCLES (FIG. 37-2)			
Superficial muscles			
Rectus abdominis	Compresses abdomen; flexes vertebral column; depresses thorax	Pubis	Upper ribs and sternum
Latissimus dorsi	Adducts, medially rotates and extends arm	Aponeuroses along middorsal line (lumbodorsal fascia)	Humerus
External oblique	Flexes and rotates vertebral column; compresses abdomen	Lumbodorsal fascia, posterior ribs	Aponeuroses along midventral surface (linea alba)
Deep muscles			
Internal oblique	Flexes and rotates vertebral column; compresses abdomen		
Transversus abdominis	Compresses abdomen	Lower ribs and lumbar vertebrae	Linea alba

142

FIG. 37-2 Ventral muscles of the thorax **(A)** and abdomen **(B)**.

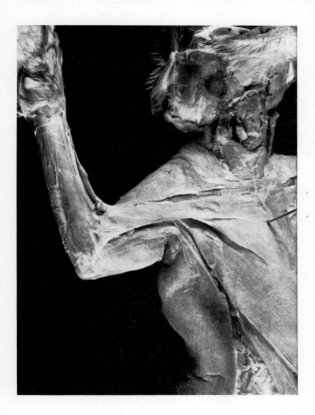

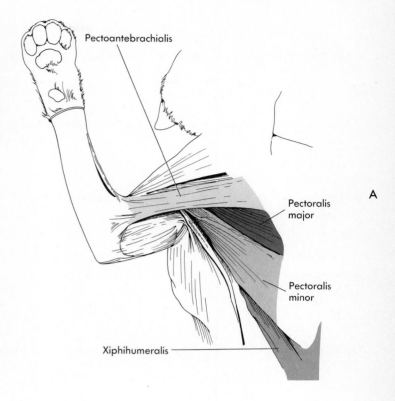

Pectoantebrachialis

Pectoralis major

Pectoralis minor

Xiphihumeralis

A

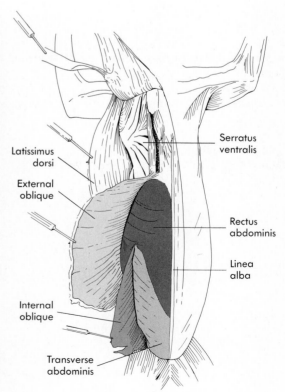

Latissimus dorsi

External oblique

Internal oblique

Transverse abdominis

Serratus ventralis

Rectus abdominis

Linea alba

B

1. What movement will occur when one sternocleidomastoid muscle is contracted?

2. What may occur when the external oblique muscle loses tone?

3. Which muscle is relaxed as the diaphragm is contracted?

4. What muscles are used to compress the abdominal cavity?

5. Which muscle maintains an erect posture of the human body trunk?

6. Explain the function of the diaphragm.

7. Name the origin and insertion of each of the following muscles:
   Cleidomastoid
   External oblique
   Rectus abdominis
   Sacrospinalis
   Iliopsoas
   External intercostals
   Scalenus
   Latissimus dorsi

*Use all references and materials at your disposal to answer these review questions.

MUSCULAR SYSTEM
Shoulder, thorax, and lumbar regions (dorsal view)

38

OBJECTIVES

1 Identify the shoulder and thoracic (dorsal view) muscles.
2 Describe the insertion and origin of the shoulder and thoracic (dorsal view) muscles.
3 Explain the function of the shoulder and thoracic (dorsal view) muscles.

KEY TERMS

Acromiodeltoid
Acromiotrapezius
Clavodeltoid
Clavotrapezius
Extensor dorsi communis
Iliocostalis
Infraspinatus
Insertion
Latissimus dorsi
Levator scapulae ventralis
Longissimus
Multifundus
Origin
Rhomboideus capitis
Rhomboideus major
Rhomboideus minor
Semispinalis
Spinalis dorsi
Spinodeltoid
Spinotrapezius
Splenius
Supraspinatus
Teres major
Teres minor

MATERIALS

dissecting animal   dissecting equipment

Most muscles join articulating bones. The muscle tendon (a cord of connective tissue) attached to the stationary bone during contraction is called the **origin,** and the muscle tendon attached to the movable bone is called the **insertion.**

The trapezius and deltoid muscles are single muscles in humans, but they are found in groups of three in the cat (Fig. 38-1). Shoulder, thorax and lumbar regions are included in Table 38-1. It is important to note that some muscles overlap other muscles.

PROCEDURE A   SUPERFICIAL MUSCLES OF THE SHOULDER AND THORAX (DORSAL VIEW)

1. Locate the muscles of the shoulder and thorax (dorsal view).
2. Describe the origin and insertion of each muscle.
3. Discuss the function of each muscle.

PROCEDURE B   DEEP MUSCLES OF THE SHOULDER AND THORAX (DORSAL VIEW)

1. Transect the clavotrapezius, acromiotrapezius, spinotrapezius, acromiodeltoid, and spinodeltoid muscles to explore the deep muscles.
2. Identify the deep muscles of the shoulder and thorax (dorsal view).
3. Discuss the function of each mucle.

PROCEDURE C   DEEP MUSCLES OF THE LUMBAR AND SACRAL REGIONS (DORSAL VIEW)

1. Transect the spinotrapezius and latissimus dorsi muscles; remove the lumbodorsal fascia. This will reveal the three extensor dorsi communis muscles, the splenius, and the multifidus.
2. Describe the origin and insertion of each muscle.
3. Discuss the function of each muscle.

FIG. 38-1 A, Dorsal superficial muscles of shoulder and thorax. **B,** Dorsal deep muscles of shoulder and thorax.

A

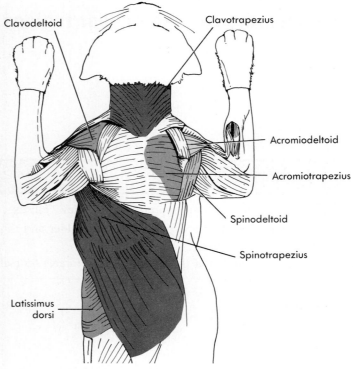

Clavodeltoid

Clavotrapezius

Acromiodeltoid

Acromiotrapezius

Spinodeltoid

Spinotrapezius

Latissimus dorsi

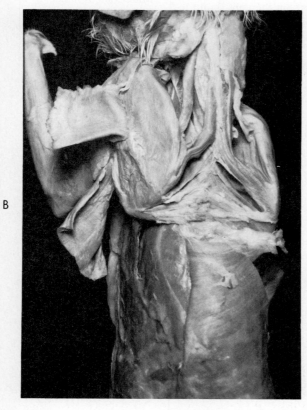

B

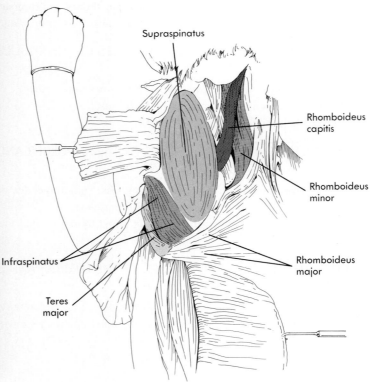

Supraspinatus

Rhomboideus capitis

Rhomboideus minor

Rhomboideus major

Infraspinatus

Teres major

TABLE 38-1 MUSCLES OF THE SHOULDER, THORAX, AND LUMBAR REGION

MUSCLE	FUNCTION	ORIGIN	INSERTION
SHOULDER AND THORAX REGION			
Superficial muscles			
Clavobrachialis or clavo-deltoid (continuous with clavotrapezius)	Extends humerus, turns head, and flexes elbow	Clavicle	Proximal end of ulna
Acromiodeltoid	Adducts and rotates humerus	Acromion process of scapula	Proximal end of humerus
Apinodeltoid	Adducts and rotates humerus	Spine of scapula	Humerus
Levator scapulae ventralis (not present in humans)	Draws scapula anteriorly	Atlas and occipital	Scapulae
Clavotrapezius (continuous with clavodeltoid)	Moves pectoral girdle and head	Back of head and middorsal line of neck	Clavicle
Acromiotrapezius	Moves pectoral girdle and head	Posterior cervical and thoracic vertebrae	Spine of scapula
Spinotrapezius	Moves pectoral girdle and head	Spinous process of posterior thoracic vertebrae	Spine of scapula
Latissimus dorsi	Adducts, medially rotates, and extends arm	Aponeuroses along middorsal line (lumbodorsal fascia)	Ventral surface of humerus
Deep muscles			
Supraspinatus	Extends scapula	Supraspinatus fossa of scapula	Greater tubercle of humerus
Infraspinatus	Rotates humerus laterally	Infraspinatus fossa of scapula	Greater tubercle of humerus
Teres major	Rotates and draws humerus posteriorly	Lateral border of scapula	Inserts in common with latissimus dorsi on proximal end of humerus
Teres minor	Assists infraspinatus rotation	Axillary border of scapula	Greater tuberosity
Rhomboideus major	Draws scapula toward vertebral column and forward	Thoracic vertebrae	Inferior angle of scapula
Rhomboideus minor	Draws scapula toward vertebral column and forward	Lower cervical and thoracic vertebrae	Dorsal border of scapula
Rhomboideus capitis	Draws scapula toward vertebral column and forward	Occipital	Vertebral border of scapula
LUMBAR REGION			
Deep muscles			
Extensor dorsi communis (sacrospinalis in humans); includes:	Extends spine; draws ribs posteriorly; bends neck and spinal column to one side	Ilium	Skull
Iliocostalis	Extends spine; draws ribs posteriorly; bends neck and spinal column to one side	Ilium; lumbar vertebrae; ribs	Ribs and transverse process of lumbar region
Longissimus (dorsi, cervicis, and capitis)	Extends spine; draws ribs posteriorly; bends neck and spinal column to one side	Processes of vertebrae	Skull (mastoid process of temporal bone)
Spinalis dorsi	Extends spine; draws ribs posteriorly; bends neck and spinal column to one side		Spinous process of upper lumbar, thoracic, and cervical vertebrae
Splenius	Extends head; flexes head laterally	First two thoracic vertebrae; dorsal midline of neck	Occipital
Multifidus	Erects and rotates vertebral column	Sacrum	Skull
Semispinalis (deep to splenis)	Extends neck and vertebral column	Vertebrae	Neck area and skull

1. Which muscle is used in shrugging the shoulders?

2. Which muscle abducts and rotates the shoulder upward?

3. What muscle is used to flex the upper arm?

4. Which upper extremity muscle is used for giving a shot?

5. Name the insertion and origin of the following muscles:
   Clavobrachialis
   Levator scapulae ventralis
   Spinotrapezius
   Latissimus dorsi
   Extensor dorsi communis
   Splenius

6. Explain the function of the following muscles:
   Clavobrachialis
   Acromiodeltoid
   Splenodeltoid
   Levator scapulae
   Spinotrapezius
   Latissimus dorsi
   Teres major
   Rhomboideus major
   Extensor dorsi communis
   Splenius

*Use all references and materials at your disposal to answer these review questions.

MUSCULAR SYSTEM
Forelimb region

39

OBJECTIVES

1 Identify the muscles of the forelimb.
2 Describe the origin and insertion of the muscles of the forelimb.
3 Explain the function of the muscles of the forelimb.

MATERIALS

dissecting animal   dissecting equipment

Most muscles join articulating bones. The muscle tendon (a cord of connective tissue) attached to the stationary bone during contraction is called the **origin,** and the muscle tendon attached to the movable bone is called the **insertion.** The forelimb region muscles are included in Table 39-1. It is important to note that some muscles overlap other muscles.

KEY TERMS

Biceps brachii
Brachialis
Brachioradialis
Epitrochlearis
Extensor carpi radialis longus
Extensor carpi ulnaris
Extensor digitorum communis
Extensor digitorum lateralis
Extensor indicis proprius
Extensor pollicis brevis
Flexor carpi ulnaris
Flexor carpi radialis
Flexor digitorum profundus
Insertion
Origin
Palmaris longus
Pronator teres
Supinator
Triceps brachii

FIG. 39-1 Forelimb. **A,** Medial, superficial view. *Continued.*

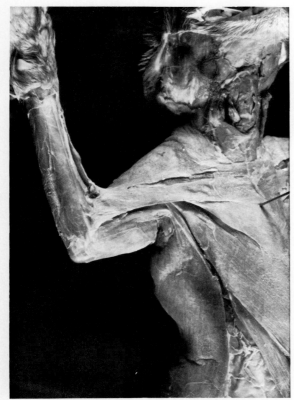

A

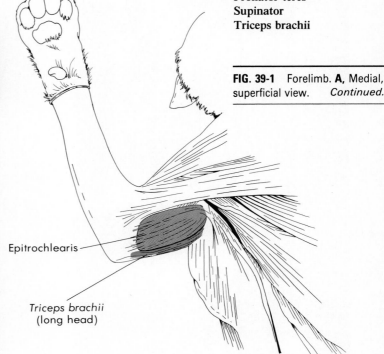

Epitrochlearis

Triceps brachii
(long head)

TABLE 39-1 MUSCLES OF THE FORELIMB

MUSCLE	FUNCTION	ORIGIN	INSERTION
MEDIAL MUSCLES (FIG. 39-1)			
Superficial muscles			
Epitrochlearis (not present in humans)	Extends elbow	Ventral border of latissimus dorsi	Distal forelimb
Deep muscles			
Biceps brachii	Flexes lower forelimb	Dorsal edge of glenoid cavity	Radial tuberosity
Triceps brachii (medial head)	Extends lower forelimb	Dorsal surface of humerus	Olecranon process of ulna
Triceps brachii (long head)	Extends lower forelimb	Scapula	Olecranon process of ulna
Brachioradialis	Rotates radius; supinates paw	Distal lateral end of humerus	Distal end (stylus) of radius
Pronator teres	Supinates lower forelimb	Medial epicondyle of humerus	Medial surface of radius
Flexor carpi radialis	Abducts and flexes wrist	Humerus	Carpals/digits
Flexor digitorum profundus	Flexes wrist and digits	Humerus	Carpals/digits
Palmaris longus	Flexes wrist	Humerus	Carpals/digits
Flexor carpi ulnaris	Flexes and adducts wrist	Humerus	Carpals/digits

FIG. 39-1, cont'd   B, Medial, deep view.

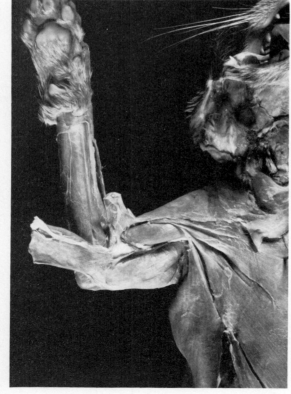

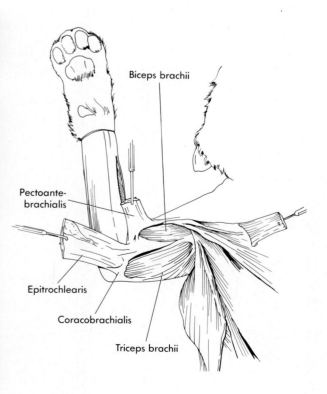

MUSCLE	FUNCTION	ORIGIN	INSERTION

LATERAL MUSCLES (FIG. 39-2)

Superficial muscles

MUSCLE	FUNCTION	ORIGIN	INSERTION
Brachialis	Flexes forelimb	Lateral surface of humerus	Proximal end of ulna
Extensor carpi radialis longus	Extends forefoot	Humerus	#2,3 metacarpal
Extensor digitorum communis	Extends digits	Humerus	#3,4,5 digits
Extensor digitorum lateralis	Extends digits	Humerus	#3,4,5 digits
Extensor carpi ulnaris	Extends wrists	Humerus	#5 metacarpal

Deep muscles

MUSCLE	FUNCTION	ORIGIN	INSERTION
Extensor pollicis brevis	Extends and abducts thumb	Ulna	#1 metacarpal
Extensor indicis proprius	Extends digit; assists extension of thumb	Ulna	#2 digit
Supinator	Supinates paw	Humerus; proximal radius	Ventral surface of radius

FIG. 39-1, cont'd C, Lateral, superficial view.

C

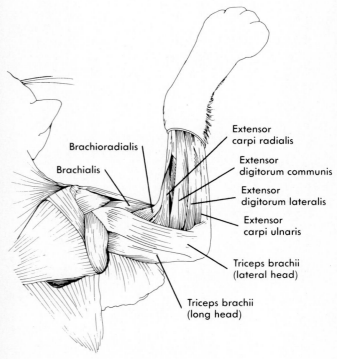

Brachioradialis

Brachialis

Extensor carpi radialis

Extensor digitorum communis

Extensor digitorum lateralis

Extensor carpi ulnaris

Triceps brachii (lateral head)

Triceps brachii (long head)

PROCEDURE A   MEDIAL MUSCLES OF THE FORELIMB

1. Examine the superficial muscles of the medial forelimb.
2. Describe the insertion and origin of each muscle.
3. Explain the function of each muscle.
4. Transect the clavobrachialis, pectoantebrachialis, pectoralis major, pectoralis minor, and epitochlearis muscles to identify the biceps and triceps muscles.
5. Transect the flexor carpi ulnaris, palmaris longus, and flexor carpi radialis muscles to examine the flexor digitorum profundus muscles.

PROCEDURE B   LATERAL MUSCLES OF THE FORELIMB

1. Identify the superficial muscles of the lateral forelimb.
2. Describe the insertion and origin of each muscle.
3. Explain the function of each muscle.
4. Transect the triceps brachii (lateral head), extensor digitorum communis, and extensor digitorum lateralis muscles to examine the extensor pollicis brevis, extensor indicis proprius, and supinator muscles.

1. List the function of the following muscles:

   Epitrochlearis

   Brachioradialis

   Pronator teres

   Flexor carpi radialis

   Flexor digitorum profundus

   Palmaris longus

   Flexor carpi ulnaris

   Brachialis

   Extensor carpi radialis longus

   Extensor digitorum communis

   Extensor digitorum lateralis

   Extensor carpi ulnaris

   Extensor pollicis brevis

   Extensor indicis proprius

   Supinator

*Use all references and materials at your disposal to answer these review questions.

40 MUSCULAR SYSTEM
Hindlimb region

KEY TERMS

Adductor femoris
Adductor longus
Biceps femoris
Caudofemoralis
Flexor digitorum longus
Gluteus maximus
Gluteus medius
Gracilis
Gastrocnemius
Iliopsoas
Insertion
Origin
Pectineus
Rectus femoris
Sartorius
Semimembranosus
Semitendinosus
Soleus
Tensor fasciae latae
Tibialis anterior
Vastus intermedius
Vastus lateralis
Vastus medialis

OBJECTIVES

1 Identify the superficial and deep muscles of the hindlimb.
2 Describe the origin and insertion of the muscles of the hindlimb.
3 Explain the function of the muscles of the hindlimb.

MATERIALS

dissecting animal   dissecting equipment

Most muscles join bones. The muscle tendon (a cord of connective tissue) attached to the stationary bone during contraction is called the *origin*, and the muscle tendon attached to the movable bone is called the *insertion*. The hindlimb muscles are included in Table 40-1. It is important to note that some muscles overlap other muscles.

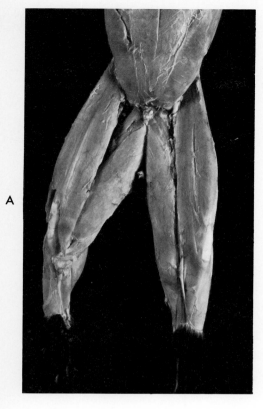

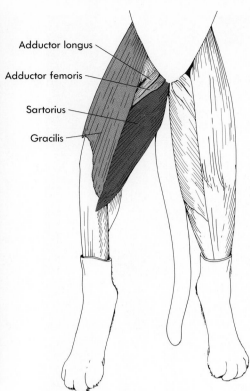

Adductor longus

Adductor femoris

Sartorius

Gracilis

A

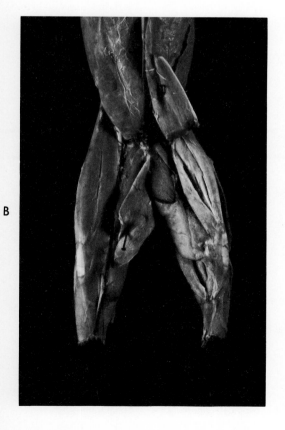

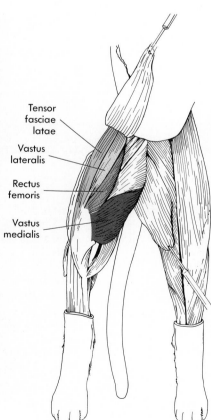

Tensor fasciae latae

Vastus lateralis

Rectus femoris

Vastus medialis

B

FIG. 40-2 Hindlimb. **A,** Dorsal view.

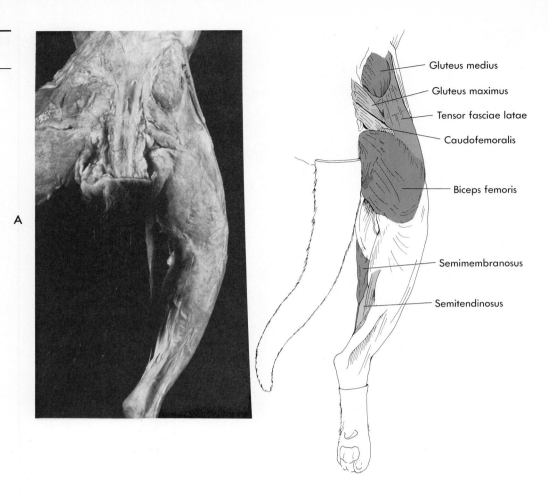

A

- Gluteus medius
- Gluteus maximus
- Tensor fasciae latae
- Caudofemoralis
- Biceps femoris
- Semimembranosus
- Semitendinosus

TABLE 40-1 MUSCLES OF THE HINDLIMB

MUSCLE	FUNCTION	ORIGIN	INSERTION
MEDIAL MUSCLES (FIG. 40-1)			
Superficial muscles			
Sartorius	Adducts and rotates thigh; extends knee	Crest of ilium	Tibia and patella
Gracilis	Adducts thigh	Pubis symphisis	Medial surface of tibia
Adductor femoris	Adducts thigh	Pubis	Femur
Adductor longus	Adducts thigh	Pubis	Femur
Pectineus	Adducts thigh	Anterior pubis	Shaft of femur
Iliopsoas (iliacus and psoas major in human)	Flexes trunk	Lumbar vertebrae and ilium	Femur
Gastrocnemius	Extends foot	Lateral and medial head of femur	Calcaneus (via Achilles tendon)
Tibialis anterior	Extends and inverts foot	Tibia and fibula	#1 metatarsal
Flexor digitorum longus	Flexes digits	Tibia and fibula	Terminal phalanges
Deep muscles			
Rectus femoris	Extends leg; flexes thigh	Iliac spine and just above acetabulum	Tibial tuberosity
Vastus lateralis	Extends leg	Lateral and dorsal surface of femur; greater trochanter	Tibial tuberosity

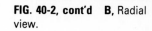

FIG. 40-2, cont'd   B, Radial view.

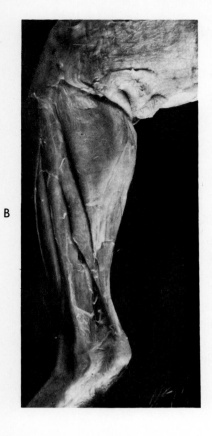

B

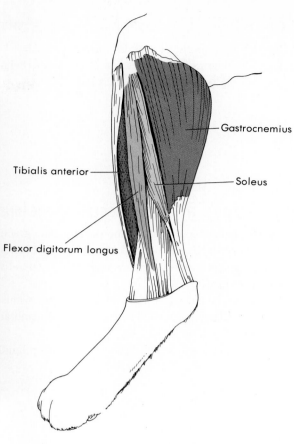

Gastrocnemius

Tibialis anterior

Soleus

Flexor digitorum longus

MUSCLE	FUNCTION	ORIGIN	INSERTION
Deep Muscles—cont'd			
Vastus medialis	Extends leg	Dorsal surface of femur	Tibial tuberosity
Vastus intermedius	Extends leg	Ventral surface of femur	Capsule of knee bone
LATERAL MUSCLES (FIG. 40-2)			
Superficial muscles			
Tensor fasciae latae	Extends thigh	Ilium	Fascia lata
Gluteus medius	Abducts thigh	Ilium and transverse processes of sacral and caudal vertebrae	Greater trochanter of femur
Gluteus maximus	Abducts thigh	Ilium and transverse processes of sacral and caudal vertebrae	Proximal femur (gluteal tuberosity)
Caudofemoralis	Abducts thigh; extends leg	Caudal vertebrae	Patella
Biceps femoralis (hamstrings)	Abducts thigh; extends leg	Ischium	Tibia and fibula
Semitendinosus (hamstrings)	Flexes knee	Ischium	Tibia
Semimembranosus (hamstrings)	Extends thigh	Ischium	Distal end of femur

PROCEDURE A SUPERFICIAL MUSCLES OF THE HINDLIMB

1. Identify the superficial muscles of the hindlimb.
2. Describe the origin and insertion of each superficial hindlimb muscle.
3. Explain the function of each muscle.

PROCEDURE B DEEP MUSCLES OF THE HINDLIMB

1. On the medial side, transect the sartorius and gracilis muscles to more thoroughly investigate the pectineus, adductor longus, adductor femoris, semimembranosus, semitendinosus, tensor fasciae latae, vastus lateralis, rectus femoris, and vastus medialis muscles.
2. On the lateral side, transect the biceps femoris muscle to more thoroughly investigate the vastus lateralis, adductor femoris, semimembranosus, and adductor femoris muscles.
3. Describe the origin and insertion of each deep hindlimb muscle.
4. Explain the function of each muscle.

1. Which muscle extends the lower leg?

2. Which muscle flexes the thigh and body trunk?

3. Abduction of the thigh is done by which two muscles?

4. Which muscles adduct the thigh?

5. Inversion of the foot would be performed by what muscle?

6. What muscle would be used to extend the foot?

7. Which muscles are involved in the extension of the leg?

8. Flexing the leg is done by which muscle group?

9. Name the origin and insertion of the following muscles:
    Gluteus maximus
    Rectus femoris
    Tensor fasciae latae
    Adductor longus
    Biceps femoris
    Gastrocnemius
    Soleus
    Vastus lateralis

*Use all references and materials at your disposal to answer these review questions.

41 MUSCULAR SYSTEM
Kinesiology

KEY TERMS

Agonist
Antagonist
Synergist

OBJECTIVES

1　Demonstrate the role of groups of muscles in body movements.
2　Describe the roles of agonists, antagonists, and synergists.

MATERIALS

laboratory partner

To provide smooth, well-coordinated movement, several skeletal muscles act in groups. **Agonists** or **prime movers** perform the desired muscular action. At the same time muscles that assist the agonist are called **synergists.** While the agonist is performing the movement, another muscle, called the **antagonist,** is relaxing. For example, as the forearm is flexed, the biceps brachii is the agonist and the triceps brachii is the antagonist. When the forearm is extended, the triceps brachii is the agonist and the biceps brachii is the antagonist.

ACTION	AGONIST	ANTAGONIST
Flexing elbow	————————	————————
Extending elbow	————————	————————
Flexing arm	————————	————————
Extending arm	————————	————————
Abducting arm	————————	————————
Adducting arm	————————	————————
Flexing knee	————————	————————
Extending knee	————————	————————
Flexing thigh	————————	————————
Extending thigh	————————	————————
Abducting leg	————————	————————
Adducting leg	————————	————————
Raising shoulder	————————	————————
Lowering shoulder	————————	————————
Turning head to side	————————	————————
Rotating thigh outward	————————	————————
Rotating thigh inward	————————	————————
Raising heel	————————	————————

PROCEDURE A MUSCLE ACTION

1. Perform the following actions and determine the agonists and antagonists.

1. Define the following terms:
   Agonist
   Synergist
   Antagonist

2. Explain the muscle action when the following muscles are agonists:
   Biceps femoris
   Rectus femoris
   Pectoralis major
   Latissimus dorsi
   Deltoid
   Sternocleidomastoid
   Flexor carpi radialis
   Extensor digitorum
   Rectus abdominis
   Psoas major
   Gluteus maximus
   Quadriceps femoris
   Hamstring
   Sartorius

*Use all references and materials at your disposal to answer these review questions.

42   DIRECTIONAL TERMS

KEY TERMS

Abdominal cavity
Abdominopelvic cavity
Anterior
Caudal
Cephalic
Cranial cavity
Deep
Distal
Dorsal
Dorsal body cavity
Epigastric region
Frontal plane
Hypochondriac region
Hypogastric
Iliac region
Inferior
Lateral
Lumbar region
Medial
Mediastinum
Midsagittal plane
Parasagittal plane
Pelvic cavity
Pericardial cavity
Pleural cavity
Posterior
Proximal
Superficial
Superior
Thoracic cavity
Transverse plane
Umbilical region
Ventral
Ventral cavity
Vertebral cavity

OBJECTIVES

1 Identify and define anatomical directional terms.
2 Locate and explain the function of the body cavities.
3 Dividing the body into planes, identify the anatomical relationship of one part to another.
4 Locate the regions of the abdominopelvic cavity.

MATERIALS

human body model   dissecting equipment
dissecting animal

   Directional terms are used to explain exactly where body structures are anatomically located. The classification of anatomical terms is associated with the anatomical position (Fig. 42-1).

Anterior or cranial	Front end or leading area of animal
Posterior	Back or trailing end of animal
Dorsal	Upper surface of animal; in humans, *dorsal* and *posterior* are synonymous
Ventral	Lower surface of animal; in humans, *ventral* and *anterior* are synonymous
Superior	Upper region of body
Inferior	Lower region of body
Cephalic	Head or brain region (used in describing four-legged animals)
Caudal	Tail region (used in describing four-legged animals)
Medial	Toward midline of body
Lateral	Toward sides of body
Proximal	Area nearest point of origin of part
Distal	Area farthest from point of origin of part
Superficial	Close to surface
Deep	Farther from surface

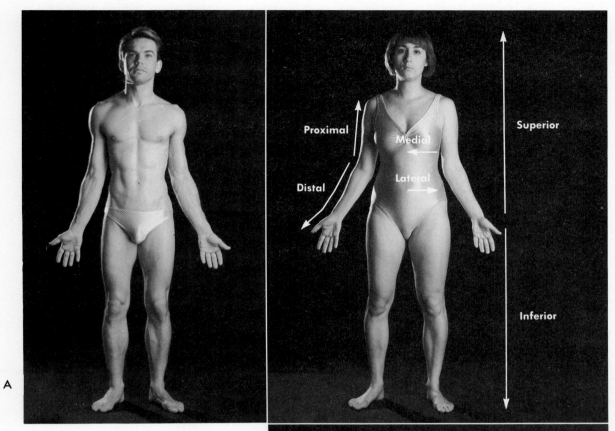

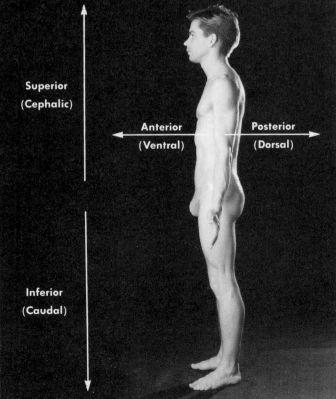

FIG. 42-1 A, Anatomical position. **B,** Directional terms from front. **C,** Directional terms from side.

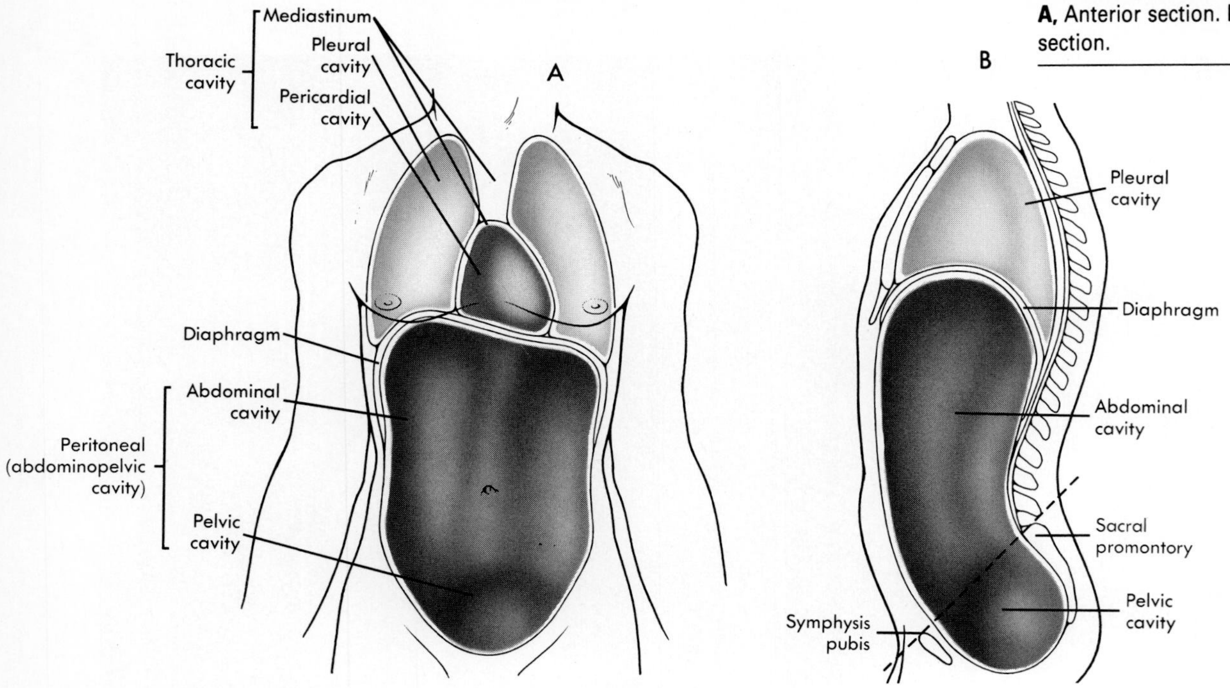

FIG. 42-2 Body cavities.
A, Anterior section. B, Sagittal section.

Body cavities are spaces in the body that contain internal organs. The classification of body cavities is as follows (Fig. 42-2):

Dorsal body cavity	Located near posterior (dorsal) surface of body
Cranial cavity	Contains brain encased within cavity of cranial bone
Vertebral cavity	Contains spinal cord and beginnings of spinal nerves; they are encased in body cavity formed by vertebrae
Ventral cavity	Located on anterior (ventral) area of body; organs within this cavity are called **viscera**
Thoracic cavity	Ventral cavity above muscular diaphragm
Pleural cavities	Each contains a lung
Pericardial cavity	Encloses heart and is located in the **mediastinum** (mass of tissue between lungs)
Abdominopelvic cavity	Ventral cavity below muscular diaphragm
Abdominal cavity	Contains most of digestive organs and some excretory organs
Pelvic cavity	Contains lower excretory organs and reproductive organs; this cavity is surrounded by bony pelvis

To more specifically identify the location of body organs, the abdominopelvic cavity may be further divided into nine regions (Fig. 42-3).

When examining body structures, **planes** may be used to identify the anatomical relationship of one part to another (Fig. 42-4).

Midsagittal plane	Divides body into equal right and left halves
Parasagittal (sagittal) plane	Any plane parallel to midsagittal plane
Frontal (coronal) plane	Divides body into anterior and posterior sections
Transverse (horizontal) plane	Divides body into superior and inferior sections

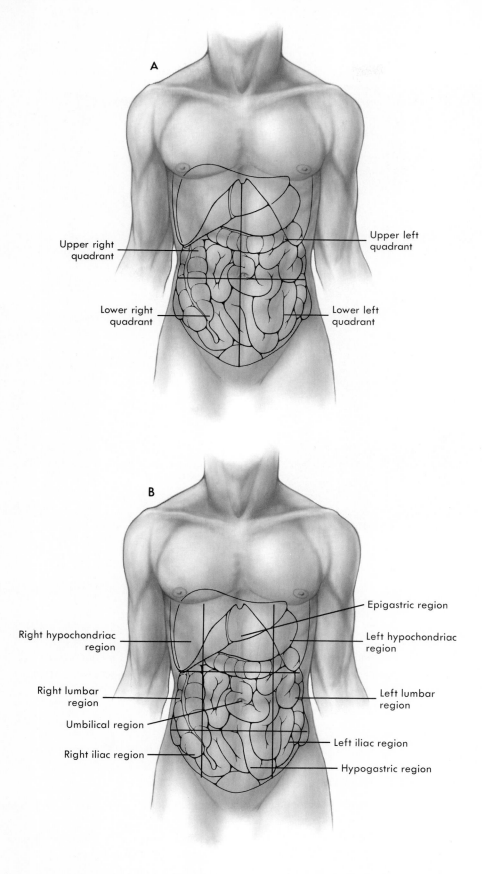

FIG. 42-3 Abdominopelvic cavity divided into nine regions.

A

Upper right quadrant

Upper left quadrant

Lower right quadrant

Lower left quadrant

B

Right hypochondriac region

Epigastric region

Left hypochondriac region

Right lumbar region

Left lumbar region

Umbilical region

Left iliac region

Right iliac region

Hypogastric region

FIG. 42-4 Planes. **A,** Body
planes. **B,** Single organ planes.

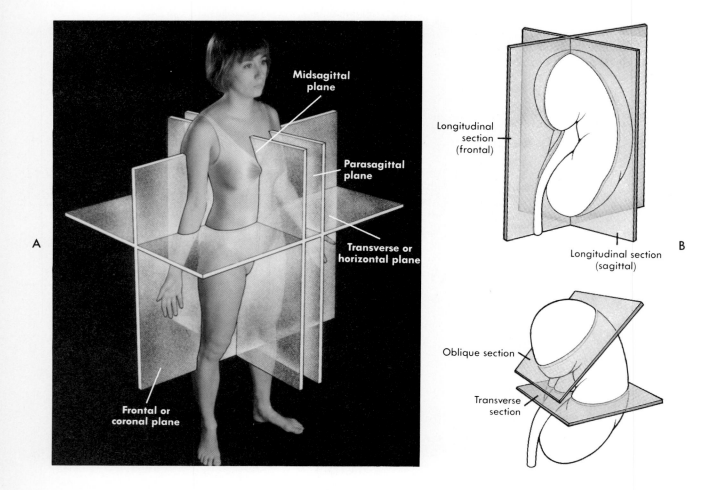

Midsagittal
plane

Parasagittal
plane

Transverse or
horizontal plane

Frontal or
coronal plane

A

Longitudinal
section
(frontal)

Longitudinal section
(sagittal)

B

Oblique section

Transverse
section

PROCEDURE A DIRECTIONAL TERMS

1. Use the human model to anatomically identify and compare each directional term.
2. Use the dissecting animal to identify and compare each directional term.

PROCEDURE B BODY CAVITIES

1. Use the human model to anatomically identify and explain the function of each body cavity.
2. About 6 mm to the left or right of the midline, make a longitudinal incision through the thoracic cavity (cutting the costal cartilage) and the abdominal cavity.
3. Cut each rib along the dorsal origin to completely expose the thoracic cavity.
4. Identify the pleural and pericardial cavities.

PROCEDURE C ABDOMINOPELVIC CAVITY

1. Use the human model to anatomically locate and identify the abdominopelvic cavity regions.
2. Locate the diaphragm, dividing the thoracic and abdominal cavities.
3. In the abdominal cavity, remove the greater omentum covering the intestines.
4. Use the dissecting animal to locate and identify the nine abdominopelvic cavity regions.

PROCEDURE D ANATOMICAL PLANES

1. Use the human model to anatomically demonstrate the planes of the body.

REVIEW QUESTIONS*

1. Compare various anatomical parts by using the following directional terms:

Anterior

Posterior

Dorsal

Ventral

Superior

Inferior

Superficial

Cephalic

Caudal

Medial

Lateral

Proximal

Distal

Deep

*Use all references and materials at your disposal to answer these review questions.

2. Explain the function of the following terms:

Cranial cavity

Vertebral cavity

Pleural cavity

Pericardial cavity

Abdominal cavity

Pelvic cavity

3. Using the following diagram of the human body, draw lines to indicate the following planes:
Midsagittal
Sagittal
Frontal
Transverse

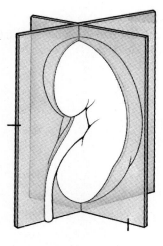

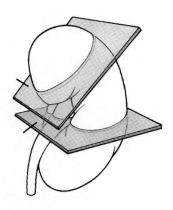

NERVOUS SYSTEM
Histology

43

OBJECTIVES

1 Identify the structure of a multipolar neuron.
2 Compare a multipolar, a bipolar, and a unipolar neuron.
3 Explain the function of a nerve.

MATERIALS

compound microscope   nerve cell slides

Neurons (nerve cells) (Fig. 43-1) function in stimulating and transmitting impulses. Structurally a neuron consists of the **nerve cell body** with short, often branched cytoplasmic extensions called **dendrites** and a long cytoplasm extension called an **axon.** Dendrites conduct an impulse toward the cell body. An axon conducts an impulse away from the cell body.

KEY TERMS

Astrocytes
Axon
Bipolar neurons
Blood-brain barrier
Dendrites
Microglia
Multipolar neurons
Myelin sheath
Nerve cell body
Neuroglia
Neurons
Oligodendrocytes
Schwann cells
Unipolar neurons

FIG. 43-1 Neurons. **A,** Multipolar neuron. **B,** Bipolar neuron. **C,** Unipolar neuron.

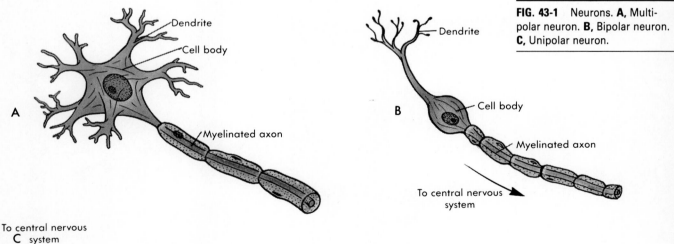

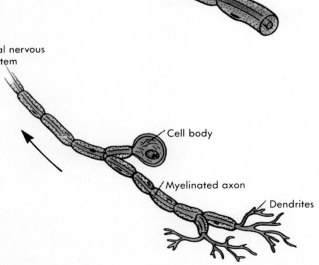

169

Neurons are classified according to their structure (Fig. 43-1 and Color Plate 11). **Multipolar neurons** have several dendrites and one axon. Most neurons in the central nervous system (brain and spinal cord) are of this type. **Bipolar neurons** have a single dendrite and a single axon. They are found in the inner ear, in the retina of the eye, and as olfactory receptors. **Unipolar neurons** have only an axon with a central branch that functions as an axon and a peripheral branch that functions as a dendrite. Most sensory neurons are unipolar.

Neuroglia (glial cells) (Fig. 43-2 and Color Plate 11) function to support and protect. **Astrocytes** are neuroglia that surround blood vessels to form the **blood-brain barrier** (protects neurons from toxic substances). **Microglia** function to engulf and destroy microbes and cellular debris. **Oligodendrocytes** are found in the central nervous system and surround axons, forming a protective **myelin sheath. Schwann cells** are found in the peripheral nervous system and also function in forming a protective myelin sheath.

PROCEDURE A IDENTIFICATION OF A MULTIPOLAR NEURON

1. Obtain a slide of a multipolar neuron; use the compound microscope to make observations on high power.
2. Draw and label a multipolar neuron.

PROCEDURE B IDENTIFICATION OF A UNIPOLAR NEURON

1. Obtain a unipolar neuron slide; use the compound microscope to make observations on high power.
2. Draw and label a unipolar neuron.

PROCEDURE C IDENTIFICATION OF NEUROGLIA

1. Obtain a neuroglial slide; use the compound microscope to make observations on high power.
2 Draw and label a multipolar neuron.

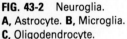

FIG. 43-2 Neuroglia.
A, Astrocyte. **B,** Microglia.
C, Oligodendrocyte.
D, Schwann cell.

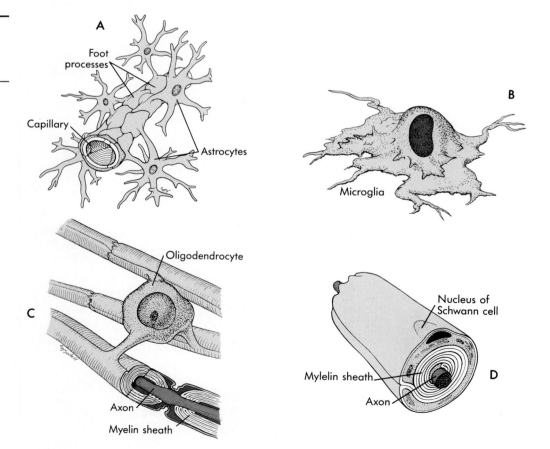

1. Draw and explain the Schwann cell stages during the formation of a myelin sheath.

2. What are neuroglia? Name the principal types and their functions.

3. Explain the chemical composition and function of a myelin sheath.

4. Define the following terms:

   Polarized membrane

   Depolarized membrane

   Repolarized membrane

5. Name and classify neurons by their shape and function.

6. Explain the all-or-none principle in impulse transmission of a neuron.

7. Define a synapse and the factors related to conduction of an impulse across a synapse.

*Use all references and materials at your disposal to answer these review questions.

NERVOUS SYSTEM
Cervical and brachial regions

OBJECTIVES

1 Identify the nerves of the cervical region.
2 Explain the function of the nerves of the cervical region.
3 Identify the nerves of the brachial region.
4 Explain the function of the nerves of the brachial region.

MATERIALS

dissecting animal   dissecting equipment

In the cat and human (mammals) there are 12 pairs of cranial nerves (see p. 24). However, the number of spinal nerves varies with the species. The cat has 38 pairs of spinal nerves. The nerves in the cat, unlike the arteries and veins, are *not* colored and are harder to identify. When locating the nerves, *care must be taken* to remove muscle, connective tissue, and fat.

The **cervical region** and **brachial region** of the nervous system include the following nerves (Figs. 44-1 and 45-1):

Hypoglossal (cranial nerve XII)	Originates from medulla; branches to some neck muscles and tongue
Spinal accessory (cranial nerve XI)	Originates from medulla and anterior spinal cord; branches to trapezius, sterno-mastoid, and cleidomastoid muscles
Vagus (cranial nerve X)	Originates from medulla; located near common carotid artery; longest of cranial nerves; supplies pharynx, larynx, lungs, heart, esophagus, stomach, and other abdominal viscera
Sympathetic trunk	Located in connective tissue beside vagus nerve; sympathetic trunk and vagus nerve lead to thorax
Spinal nerves	There are eight cervical and thirteen thoracic spinal nerves; spinal nerves pass from vertebral column through intervertebral foramen and divide to form two **rami**; **dorsal ramus** innervates dorsal body wall and skin and is smaller; **ventral ramus** is larger and forms networks or **plexuses** to innervate ventral skeletal muscles, arms, and legs
Brachial plexus	Formed by ventral rami of first thoracic nerve and last four cervical nerves; innervates muscles of limbs and thorax
Suprascapular	Originates from sixth cervical nerve; extends over anterior part of scapula; innervates supraspinatus muscle
Subscapular	Originates from sixth, seventh, and eighth cervical nerves; innervates subscapularis, teres major, and latissimus dorsi muscles

Phrenic	Fifth and sixth cervical nerves fuse together to become phrenic nerve; innervates diaphragm and is vital to respiration
Axillary	Originates from sixth and seventh cervical nerves; innervates head of triceps brachii and spinodeltoid muscles
Radial	Largest nerve of brachial plexus; innervates triceps brachii and continues to form superficial radial nerve
Median	Innervates ventral muscles of lower forelimb
Ulnar	Extends on posterior side of lower forelimb; innervates carpi ulnaris muscle
Ventral thoracic	Originates from eighth cervical and first thoracic nerves; innervates underside of pectoralis muscles
Long thoracic	Originates from seventh cervical nerve; innervates serratus ventralis

PROCEDURE A CERVICAL REGION

1. With the ventral side up, place the dissecting animal on the dissecting tray.
2. Locate the right common carotid artery just below the jaw; use a dissecting needle to *carefully* remove any muscle, connective tissue, and fat; locate the hypoglossal nerve.
3. Below the hypoglossal nerve, locate the superior cervical and nodose ganglia; from here locate the spinal accessory nerve.
4. Below the superior cervical ganglia and nodose ganglia, locate the vagus nerve and the sympathetic trunk; both nerves lie lateral to the common carotid artery.
5. Remove tissue from the vertebral region to expose the cervical spinal nerves.
6. Explain the function of each nerve.

PROCEDURE B BRACHIAL REGION

1. Locate the following nerves that extend from the brachial plexus: suprascapular, subscapular, phrenic, axillary, radial, median, ulnar, ventral thoracic, and long thoracic.
2. Explain the function of each nerve.

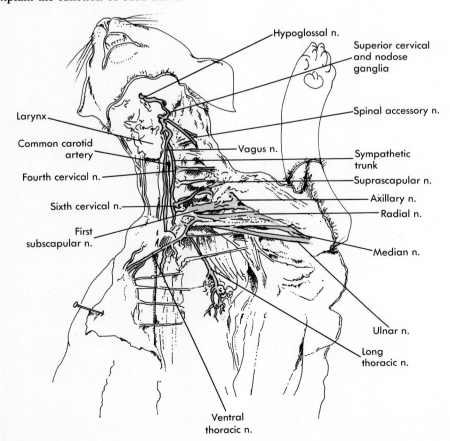

FIG. 44-1 Nervous system, cervical and brachial regions.

1. List the factors that may inhibit or block nerve impulses.

2. Define summation.

3. Describe the route the herpesvirus and rabies virus make. Where do they multiply?

4. Distinguish between sciatica and neuritis.

5. Discuss the cause and symptoms of shingles.

6. Explain what dermatomes are and where they are located.

7. What are cervical and lumbar enlargements?

8. Define a plexus.

9. Describe the composition and functions of the principal plexuses.

10. What is Wallerian degeneration?

*Use all references and materials at your disposal to answer these review questions.

NERVOUS SYSTEM
Thoracic and abdominal regions

Abdominal region
Celiac ganglia
Celiac plexus
Dorsal vagus trunk
Greater splanchnic nerve
Lesser splanchnic nerve
Recurrent laryngeal nerve
Superior mesenteric ganglia
Sympathetic nerve trunk
Thoracic region
Thoracic spinal trunk
Vagus nerve
Ventral vagus trunk

OBJECTIVES

1 Identify the nerves of the thoracic and abdominal regions.
2 Explain the function of the nerves of the thoracic and abdominal regions.

MATERIALS

dissecting animal   dissecting equipment

The **thoracic region** and **abdominal region** of the nervous system include the following nerves (Figs. 45-1 and 45-2):

Vagus nerve	**Parallel to trachea and esophagus; near subclavian artery, recurrent laryngeal nerves** branch off to innervate larynx; network branches off to innervate lungs and heart; divides into **dorsal vagus trunk** and **ventral vagus trunk** and passes through diaphragm; below diaphragm, in abdominal region, branches to innervate stomach; branches to **celiac plexus,** which innervates abdominal viscera
Sympathetic nerve trunks	Innervates heart; branches off to **greater splanchnic** and **lesser splanchnic nerves;** below diaphragm synapse with **celiac ganglia** and **superior mesenteric ganglia,** which innervate abdominal viscera
Thoracic spinal trunk	Contains 13 nerve pairs; first pair contribute to brachial plexus; remaining nerves of ventral rami innervate intercostal muscles

PROCEDURE A   IDENTIFICATION OF THORACIC AND ABDOMINAL NERVES

1. *Carefully* remove any muscle, connective tissue, and fat to identify the following thoracic and abdominal nerves:
   Phrenic nerve
   Vagus nerve
   Sympathetic nerve trunk
   Thoracic spinal nerves
   Celiac plexus
   Lesser splanchnic
   Greater splanchnic
2. As far as possible, follow the branches of each nerve to the point of innervation.
3. Explain the function of each nerve.

FIG. 45-1 Nervous system, thoracic region.

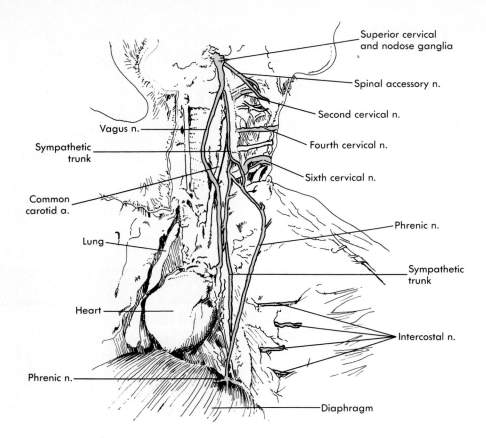

Vagus n.

Sympathetic trunk

Common carotid a.

Lung

Heart

Phrenic n.

Superior cervical and nodose ganglia

Spinal accessory n.

Second cervical n.

Fourth cervical n.

Sixth cervical n.

Phrenic n.

Sympathetic trunk

Intercostal n.

Diaphragm

FIG. 45-2 Nervous system, abdominal and lumbosacral region.

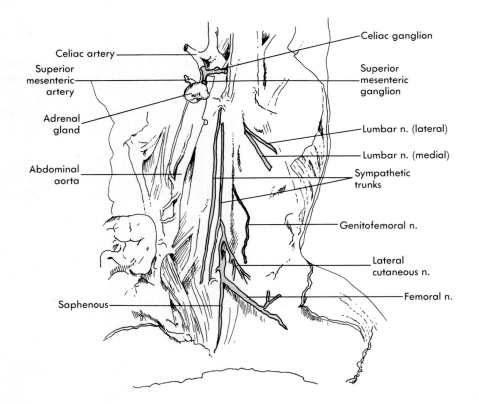

Celiac artery

Superior mesenteric artery

Adrenal gland

Abdominal aorta

Saphenous

Celiac ganglion

Superior mesenteric ganglion

Lumbar n. (lateral)

Lumbar n. (medial)

Sympathetic trunks

Genitofemoral n.

Lateral cutaneous n.

Femoral n.

1. Describe how a spinal nerve is attached to the spinal cord.

2. How are spinal nerves named and numbered?

3. What are the intercostal nerves?

4. Define the term *transection*.

5. Explain spinal shock.

6. What is the clinical importance of an abdominal reflex?

7. Where do abdominal nerves originate?

8. What is meningitis? How is it diagnosed?

9. Define the following disorders:

   Cerebral palsy

   Poliomyelitis

   Headache

   Epilepsy

*Use all references and materials at your disposal to answer these review questions.

46 NERVOUS SYSTEM
Lumbar and sacral regions

OBJECTIVES

1 Identify nerves of the lumbosacral plexus.
2 Explain the function of the lumbosacral plexus.

MATERIALS

dissecting animal   dissecting equipment

The **lumbosacral plexus** contains the last four lumbar nerves and the three sacral nerves. The nerves of the lumbosacral plexus are as follows (Fig. 46-1):

Genitofemoral	Originates from fourth and fifth lumbar nerves; innervates integument, external genitalia, medial side of thigh, and adjacent body wall
Lateral cutaneous	Originates from fourth and fifth lumbar nerves; innervates integument on lateral side of thigh
Femoral	Originates from fifth and sixth lumbar nerves; innervates ventral femoral muscles
Obturator	Originates from sixth and seventh lumbar nerves; passes posteriorly through obturator foramen; innervates femoral, adductor, pectineus, and gracilis muscles
Lumbosacral cord	Originates from sixth and seventh lumbar nerves; fuses last two lumbar nerves to sacral nerves
Sciatic nerve	Lumbosacral cord and first sacral nerve join together to form sciatic nerve; innervates thigh and lower leg muscles
Muscular branch	Branches from portion of sciatic nerve; innervates biceps femoris, semitendinosus, semimembranosus, and abductor femoris muscles
Sural	Branches from sciatic nerve; innervates gastrocnemius
Common peroneal and tibial	Branches from sciatic nerve; innervates lower leg

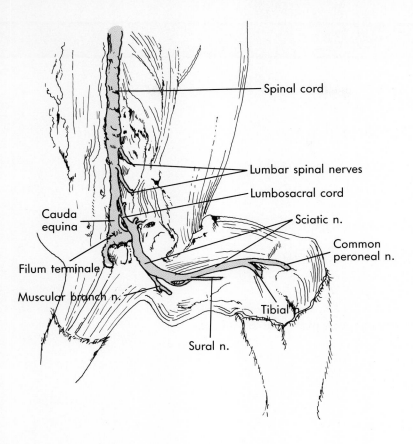

FIG. 46-1 Nervous system. Lumbosacral region (dorsal view).

Spinal cord

Lumbar spinal nerves

Lumbosacral cord

Cauda equina

Sciatic n.

Common peroneal n.

Filum terminale

Muscular branch n.

Tibial n.

Sural n.

PROCEDURE A   IDENTIFICATION OF LUMBOSACRAL NERVES

1. *Carefully* remove any muscle, connective tissue, and fat to examine the genitofemoral, lateral cutaneous, femoral, and obturator nerves; it may also be necessary to *carefully* move the kidney, horn of the uterus, and the urinary bladder.
2. Place the dorsal side of the dissecting animal on the dissecting tray.
3. Transect the biceps femoris, tensor fasciae latae, caudofemoralis, gluteus maximus, and gluteus minimus muscles to expose the lumbosacral cord and lumbosacral nerves.
4. Explain the function of each nerve.

1. Explain the symptoms of syphilis.

2. Explain the proper procedure used in performing a spinal tap.

3. Discuss what would happen if the sciatic nerve were pinched at the spine.

4. What causes excessive curvature of the spine?

5. Compare the number of each type of spinal nerve in cats and humans.

*Use all references and materials at your disposal to answer these review questions.

NERVOUS SYSTEM
Spinal cord gross anatomy

47

OBJECTIVES
1 Identify the gross anatomy of the spinal cord.
2 Explain the function of the spinal cord.

KEY TERMS

Anterior gray horns
Anterior white column
Arachnoid
Ascending tracts
Cauda equina
Central canal
Cerebrospinal fluid
Cervical enlargement
Conus medullaris
Descending tracts
Dorsal root
Dura mater
Gray commissure
Gray matter
Lateral white column
Lumbar enlargement
Meninges
Motor impulses
Pia mater
Posterior gray horns
Posterior white column
Sensory impulses
Spinal cord
Spinal ganglia
Ventral root
White matter

MATERIALS

dissecting animal   dissecting equipment

The **spinal cord** functions to carry **sensory impulses** to the brain and carry **motor impulses** away from the brain. It also functions to integrate reflexes.

The spinal cord is a continuation of the medulla oblongata and passes through the foramen magnum. Because many nerves originate in the limbs, the spinal cord diameter is thickest at the **cervical enlargement** (fourth cervical to first thoracic vertebrae) and the **lumbar enlargement** (ninth to twelfth thoracic vertebrae). The end of the spinal cord has a conical shape and is called the **conus medullaris** (first or second lumbar vertebrae) (Fig. 47-1).

Several protective layers of connective tissue, called the **meninges,** cover the spinal cord. The outer layer of the spinal meninges is called the **dura mater;** the middle layer is the **arachnoid;** and the inner layer is the **pia mater.** At the end of the conus medullaris is an extension of the meninges, mostly pia mater, called the **filum terminale** (see Fig. 47-1). Even though the spinal cord proper ends between L1 and L2, the lumbar, sacral, and coccyx spinal nerves continue down through the vertebral canal before emerging. This "fanning" pattern is referred to as the **cauda equina** (horse's tail).

The nervous system contains **white matter,** aggregations of myelinated axons and supporting neuroglia, and **gray matter,** nerve cell bodies and dendrites or bundles of un-myelinated axons and supporting neuroglia. A cross-section (Fig. 47-2) of the spinal cord reveals white matter surrounding the gray matter shaped like the letter H (or butterfly). The crossbar of the H is known as the **gray commissure.** The small space in the center of the gray commissure is called the **central canal** and contains **cerebrospinal fluid.** The anterior upright portions of the H form the motor part of the gray matter and are called the **anterior (ventral) gray horns.** The posterior upright portions of the H form the sensory part of the gray matter and are called the **posterior (dorsal) gray horns.** Exiting near the ventral and dorsal horns are the **ventral root** and the **dorsal root.** The ventral root carries impulses away from the spinal cord, and the dorsal root carries information to the spinal cord. The dorsal roots form the **dorsal root ganglia (spinal ganglia)** that contain the cell bodies of these sensory neurons (see Fig. 47-2).

The white matter is divided into the **anterior white column, posterior white column,** and **lateral white column.** Each column consists of **tracts** or bundles of myelinated fibers. Sensory axons, carrying impulses toward the brain, form the **ascending tracts.** Motor axons, carrying impulses away from the brain, form the **descending tracts.**

FIG. 47-1 Spinal cord and spinal nerves.

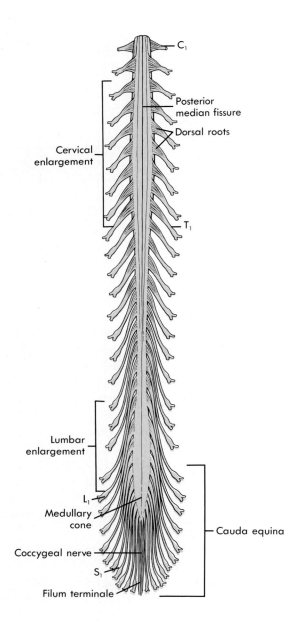

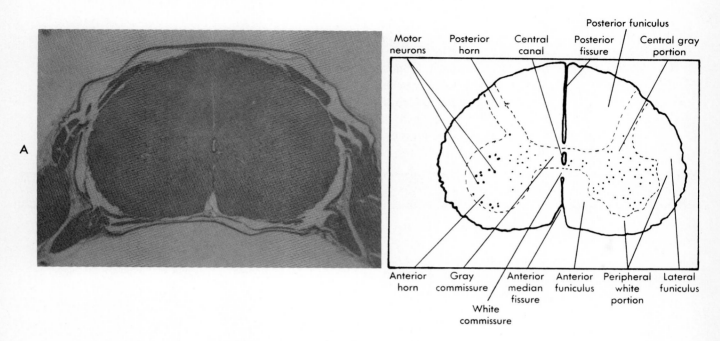

A

Motor neurons
Posterior horn
Central canal
Posterior fissure
Posterior funiculus
Central gray portion

Anterior horn
Gray commissure
White commissure
Anterior median fissure
Anterior funiculus
Peripheral white portion
Lateral funiculus

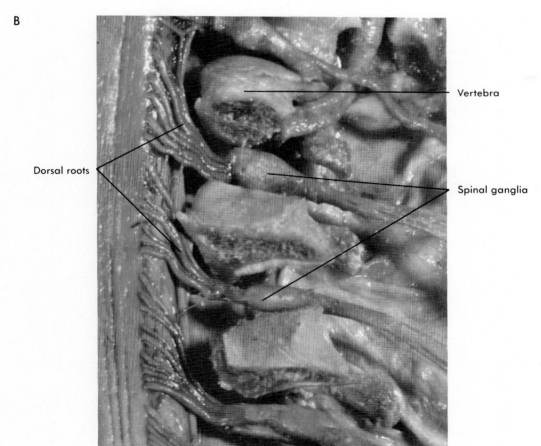

B

Dorsal roots

Vertebra

Spinal ganglia

PROCEDURE A EXPOSURE OF THE SPINAL CORD

1. Place the dissecting animal, dorsal side up, on the dissecting tray; cut away the muscles dorsal to the vertebral column.
2. To expose the spinal cord, use bone cutters or scissors to cut the neural arches of the vertebrae.

PROCEDURE B IDENTIFICATION OF A SPINAL CORD CROSS-SECTION

1. Cut a small, 5 mm section of the spinal cord and observe the cross-section under a dissecting microscope.
2. Locate the gray matter and white matter.
3. Draw and properly label the cross-section.
4. Explain the function of each structure.

REViEW QUESTIONS*

1. Describe the gross anatomy cross-section of the spinal cord.

2. Explain how the spinal cord is protected.

3. Name the 31 pairs of spinal nerves.

4. Explain the effects of injury on cranial nerves.

5. Which spinal root carries sensory information, and which spinal root carries motor information? Why are they called mixed nerves?

*Use all references and materials at your disposal to answer these review questions.

NERVOUS SYSTEM
Brain gross anatomy

48

OBJECTIVES

1 Identify the structures of the brain.
2 Explain the function of the structures of the brain.
3 Locate the cranial nerves.

MATERIALS

dissecting animal   dissecting equipment

The **brain** (Fig. 48-1) is protected by the **cranial bones** and the **cranial meninges,** an extension of the spinal meninges. The brain is highly convoluted. Depressions in the brain surface are called **sulci.** Raised areas on the brain surface are called **gyri.** Major anatomical areas of the brain are the **cerebrum, cerebellum, midbrain, pons,** and **medulla.** The cerebrum portion of the brain is divided by the **longitudinal cerebral fissure** into the right and left **cerebral hemispheres.**

Located on the ventral surface of the brain one may observe paired cranial nerves (Fig. 48-2). From anterior to posterior are the **olfactory bulb** (I), **optic nerve** (II), **occulomotor nerve** (III), **trochlear nerve** (IV), **trigeminal nerve** (V), **abducens nerve** (VI), **facial nerve** (VII), **vestibulocochlear nerve** (VIII), **glossopharyngeal nerve** (IX), **vagus nerve** (X), **accessory nerve** (XI), and **hypoglossal nerve** (XII).

In a midsagittal section of the brain, one may observe the following (Fig. 48-3):

Medulla oblongata	Regulates heart rate, respiratory rate, vasoconstriction, swallowing, coughing, vomiting, sneezing, and hiccuping
Pons varolii	Links spinal cord to brain; links parts of brain with one another; relays impulses related to voluntary skeletal movements from cerebral cortex to cerebellum; assists in respiratory regulation
Midbrain	Relays motor impulses from cerebrum to cerebellum and spinal cord; relays sensory impulses from spinal cord to thalamus; regulates visual and auditory reflexes
Cerebellum	Coordinates skeletal muscles; maintains posture and balance
Arbor vitae	"Tree of life"; branched white matter within cerebellum
Thalamus	Relay station for sensory impulses, except smell; recognition of touch, pressure, pain, and temperature
Hypothalamus	Regulates autonomic nervous system, body temperature, sleep, waking state, fluid and food intake, sex drive (as part of limbic system), and secretions (hormone release) of the pituitary
Corpus callosum	White matter, major commisural tract connecting right and left hemispheres

KEY TERMS

Abducens nerve
Accessory nerve
Arbor vitae
Association areas
Basal ganglia
Brain
Cerebellum
Cerebral aqueduct
Cerebral hemispheres
Cerebrum
Corpus callosum
Cranial meninges
Facial nerve
Fourth ventricle
Glossopharyngeal nerve
Gyri
Hypoglossal nerve
Hypothalamus
Intraventricular foramen
Lateral aperture
Lateral ventricle
Limbic system
Longitudinal cerebral fissure
Median aperture
Medulla oblongata
Midbrain
Motor areas
Oculomotor nerve
Olfactory bulb
Optic nerve
Pons varolii
Sensory areas
Sulci
Tentorium
Thalamus
Third ventricle
Trigeminal nerve
Trochlear nerve
Vagus nerve
Vestibulocochlear nerve

FIG. 48-1 Brain covering, spaces, and convolutions.

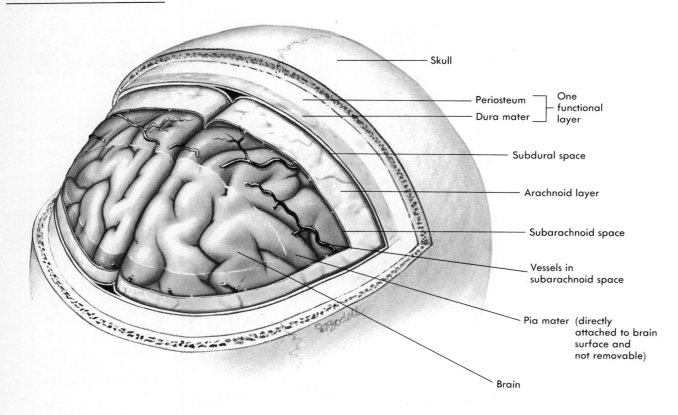

Skull

Periosteum ⎤ One
Dura mater ⎦ functional layer

Subdural space

Arachnoid layer

Subarachnoid space

Vessels in subarachnoid space

Pia mater (directly attached to brain surface and not removable)

Brain

FIG. 48-2 Cranial nerves (inferior view).

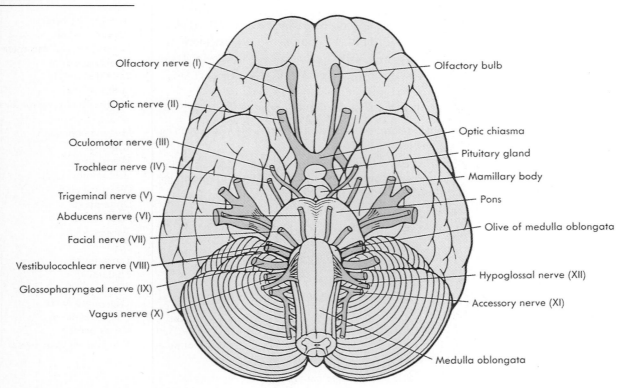

Olfactory nerve (I)

Optic nerve (II)

Oculomotor nerve (III)

Trochlear nerve (IV)

Trigeminal nerve (V)

Abducens nerve (VI)

Facial nerve (VII)

Vestibulocochlear nerve (VIII)

Glossopharyngeal nerve (IX)

Vagus nerve (X)

Olfactory bulb

Optic chiasma

Pituitary gland

Mamillary body

Pons

Olive of medulla oblongata

Hypoglossal nerve (XII)

Accessory nerve (XI)

Medulla oblongata

FIG. 48-3 Midsagittal section of the brain.

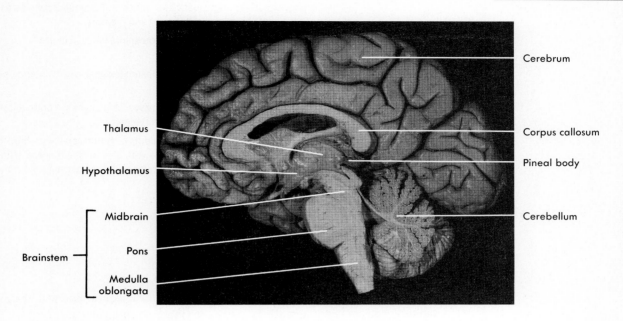

Cerebrum

Thalamus

Corpus callosum

Hypothalamus

Pineal body

Brainstem

Midbrain

Cerebellum

Pons

Medulla
oblongata

FIG. 48-4 Ventricles of the brain.

Anterior horn:
lateral ventricle

Posterior horn:
lateral ventricle

Interventricular
foramen

Third ventricle

Inferior horn:
lateral ventricle

Cerebral aqueduct

Fourth ventricle

Central canal:
spinal cord

Cerebrum	**Basal ganglia** control muscular movement; **motor areas** regulate muscular movement; **sensory areas** interpret sensory impulses; **association areas** control intellectual processes and emotion; **left hemisphere** regulates linguistics; **right hemisphere** analyzes nonverbal visual patterns and regulates melodies; **limbic system** regulates emotions of behavior and memory
Ventricles (Fig. 48-4)	Cavities within brain through which cerebrospinal fluid circulates
Lateral ventricles	One in each hemisphere below corpus callosum (extensive with anterior, posterior, and temporal bones)
Third ventricle	Between and inferior to right and left halves of thalamus; communicates with lateral ventricles via **intraventricular foramen**
Fourth ventricle	Between inferior brain stem and cerebellum; communicates with third ventricle via **cerebral aqueduct**; communicates with subarachnoid space of meninges and two **lateral apertures**; communicates with the central canal of the spinal cord via a median aperature

PROCEDURE A   REMOVAL OF THE BRAIN

1. Remove any muscle tissue around the skull.
2. Once the skull is exposed, use a bone saw to make a caplike (in line with the external auditory meatus) cut around the cranium; also make a midsagittal cut.
3. With forceps, *carefully* break away the section of bone.
4. Between the cerebrum and cerebellum is a transverse bony septum called the **tentorium** (*cerebelli,* dura mater in human); take extra care in its removal.
5. Clip the optic nerves and cut the brain away from the spinal cord.
6. Remove any meninges.

PROCEDURE B   IDENTIFICATION OF CRANIAL NERVES

1. Examine the ventral side to observe the olfactory bulb, optic nerve, oculomotor nerve, trochlear nerve, trigeminal nerve, abducens nerve, facial nerve, vestibulocochlear nerve, glossopharyngeal nerve, vagus nerve, accessory nerve, and hypoglossal nerve.

PROCEDURE C   IDENTIFICATION OF BRAIN MIDSAGITTAL SECTION

1. Identify the sulci, gyri, longitudinal cerebral fissure, cerebrum, cerebellum, medulla oblongata, and pons varolii.
2. Make a midsagittal section in the brain; identify and explain the function of the cerebrum, cerebellum, arbor vitae, medulla oblongata, pons varolii, midbrain, thalamus, hypothalamus, corpus callosum, and ventricles.

1. Explain the structure and function of the cranial meninges.

2. Explain the formation and circulation of the cerebrospinal fluid.

3. Discuss brain lateralization and the split brain concept.

4. Describe the location and explain the function of the limbic system.

5. Compare the components of the brain system with regard to structure and function.

6. Describe the structural features of the cerebrum.

7. Describe the function of the cerebellum.

8. Explain the various chemical transmitter substances found in the brain.

*Use all references and materials at your disposal to answer these review questions.

49

NERVOUS SYSTEM
Human reflexes

OBJECTIVES

1 Identify and determine the responses obtained by several common clinical reflex tests.

MATERIALS

percussion or reflex hammer   laboratory partner
metal stick

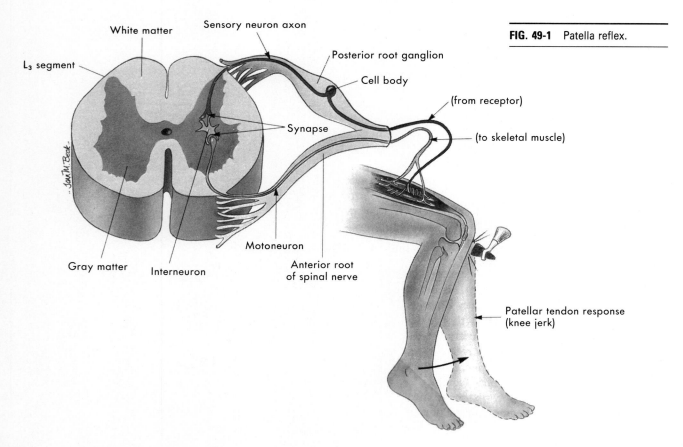

FIG. 49-1 Patella reflex.

White matter
Sensory neuron axon
Posterior root ganglion
L₃ segment
Cell body
(from receptor)
Synapse
(to skeletal muscle)
Gray matter
Interneuron
Motoneuron
Anterior root
of spinal nerve
Patellar tendon response
(knee jerk)

A light tap to a **tendon** can stimulate contraction of a skeletal muscle. The responding reflex can provide information about the reflex and spinal cord areas associated with specific muscles.

The **patellar reflex** (Fig. 49-1), or knee jerk, is performed by tapping the patellar ligament. Under normal conditions, contraction of the quadriceps femoris muscle will cause extension of the leg. Damage to the **second, third,** or **fourth lumbar segments** of the spinal cord or to the afferent and efferent nerves will block the reflex.

The **Achilles reflex** (Fig. 49-2, *A*), or ankle jerk, is performed by tapping the Achilles tendon. Under normal conditions, contraction of the gastrocnemius and soleus muscles will cause extension of the foot. Damage to the **first** or **second sacral segments** of the spinal cord or nerves of the posterior leg muscles will block the reflex.

The **plantar flexion** or **Babinski reflex** (Fig. 49-2, *B*), is performed by stimulating the outer sole of the foot. Under normal conditions extension of the foot occurs in individuals under 1½ years of age. In those over 1½ years who have **corticospinal tract** lesions, this reflex will be blocked.

The **triceps reflex** (Fig. 49-2, *C*) is performed by tapping the tendon of the triceps. Under normal conditions, extension of the forearm will occur. This indicates proper functioning of the **sixth, seventh,** and **eighth cervical segments** of the spinal cord.

The **biceps reflex** (Fig. 49-2, *D*) is performed by tapping the biceps tendon. Under normal conditions, flexion of the forearm will occur. This indicates proper functioning of the **fifth** and **sixth cervical segments** of the spinal cord.

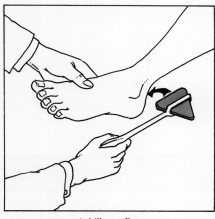

Achilles reflex

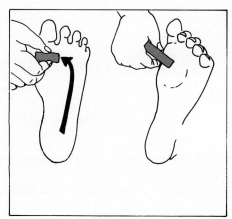

Babinski reflex

FIG. 49-2 **A,** Achilles reflex. **B,** Plantar (Babinski) reflex. **C,** Triceps reflex. **D,** Biceps reflex.

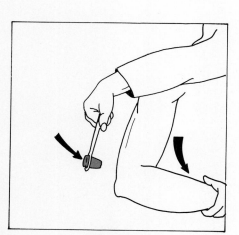

Triceps reflex

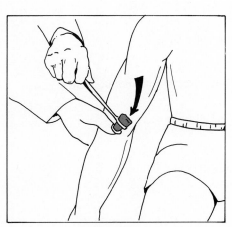

Biceps reflex

PROCEDURE A PATELLAR REFLEX

1. Have your laboratory partner sit on a table and let his or her leg hang free.
2. Using the reflex hammer, tap the patellar ligament just below the knee.
3. Identify the muscle responding to the reflex contraction.

PROCEDURE B ACHILLES REFLEX

1. Have your laboratory partner kneel on a stable chair while letting the feet hang free over the edge of the chair.
2. Using the reflex hammer, tap the Achilles tendon.
3. Identify the muscle responding to the reflex contraction.

PROCEDURE C BABINSKI REFLEX

1. Have your laboratory partner sit on a table and let the leg hang free.
2. Using a metal stick, lightly stimulate the outer margin of the sole (from the heel to the toe).
3. Note the response.

PROCEDURE D TRICEPS REFLEX

1. Place your laboratory partner's elbow in your left hand; using the reflex hammer, tap the triceps tendon.
2. Identify the muscle responding to the reflex contraction.

PROCEDURE E BICEPS REFLEX

1. Have your laboratory partner rest his or her arm on the arm of a chair; place your thumb on the biceps tendon and use the reflex hammer to tap your thumb.
2. Identify the muscle responding to this reflex contraction.

REVIEW QUESTIONS*

1. Define a reflex.

2. What are the basic components of a reflex?

3. Compare a spinal, somatic, and visceral (autonomic) reflex.

4. Draw, label, and explain a stretch reflex.

5. Draw, label, and explain a flexor reflex.

*Use all references and materials at your disposal to answer these review questions.

SPECIAL SENSES
Visual gross anatomy

50

OBJECTIVES

1 Describe the accessory structures of the eye.
2 List the structures of the eyeball.
3 Explain the function of each eyeball structure.

MATERIALS

dissecting animal eye   dissecting equipment

Accessory structures of the eye that provide protection for the eye include the following (Fig. 50-1):

Eyebrows	Protection from perspiration and direct light rays
Palpebrae (eye-lids)	Protection from foreign objects, excessive light; evenly distribute lubricating secretions; contain levator palpebrae superioris muscle

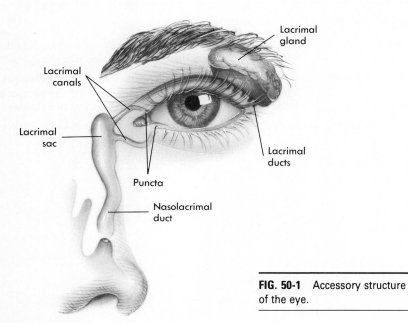

FIG. 50-1 Accessory structure of the eye.

Lateral and medial canthi (junctions of upper and lower eyelids)	Corner of the eye; medial canthus is location of **lacrimal caruncle,** containing sebaceous and sudoriferous glands
Tarsal plate	Eyelid inner wall layer of connective tissue; provides form and support for eyelid; contains **Meibomian glands;** modified sebaceous gland secretions help prevent eyelids from adhering to one another
Conjunctiva	Mucous membrane layer of eyelid and exposed surface of eyeball
Lacrimal gland	Compound **tubuloacinar gland** which excretes **lacrimal secretions** containing lubricants, wastes, salts, and **lysozyme,** a bacterial enzyme; opening of **lacrimal duct** is called **punctum**

The **eyeball** (Fig. 50-2) structure is divided into three basic layers. The outer layer is called the **fibrous tunic,** the middle layer is called the **vascular tunic,** and the inner layer is called the **retina.** Each layer of the eye is divided into the following structures:

Fibrous tunic

Sclera	"White of eye"; gives shape to eyeball and protects inner components; posterior surface receives optic nerve
Cornea	Transparent, nonvascular, fibrous coat covering iris; contains outer epithelial layer

Vascular tunic

Choroid	Dark brown to black membrane lining internal surface of sclera; highly vascular and pigmented
Ciliary body	Anterior portion of choroid; contains smooth muscles that alter shape of lens for near or far vision
Iris	Smooth muscle with hole in center called **pupil;** regulates amount of light entering eyeball

Retina

Nervous layer	Covers internal surface of choroid except at cilary body; appears to end at scalloped border known as **ora serrata;** contains photoreceptor neurons called **rods** and **cones.** Cones are specialized for color and sharpness; rods for shades, shape, and movement. Where **optic nerve** and blood vessels enter and exit is blind spot. Just lateral to the blind spot is the central fovea. This is a small depression containing densely concentrated cones; area of sharpest vision
Pigmented layer	Nonvisual portion of retina lying next to choroid; absorbs light and prevents reflection

Inside the layers of the eyeball are the "marble looking" **lens,** the watery **aqueous humor,** and the jellylike **vitreous humor.**

Lens	Flexible biconvex structure made up of successive layers of protein fibers; held in position by **suspensory ligaments,** allows for accommodation
Aqueous humor	Thought to be secreted by choroid; flows from **posterior chamber,** posterior to iris, to **anterior chamber,** anterior to the iris; maintains **intraocular pressure;** continuously secreted by choroid and drained by **canal of Schlemm**
Vitreous humor	Gel found in posterior cavity behind the lens, formed during embryonic development and not replaced; prevents eye from collapsing

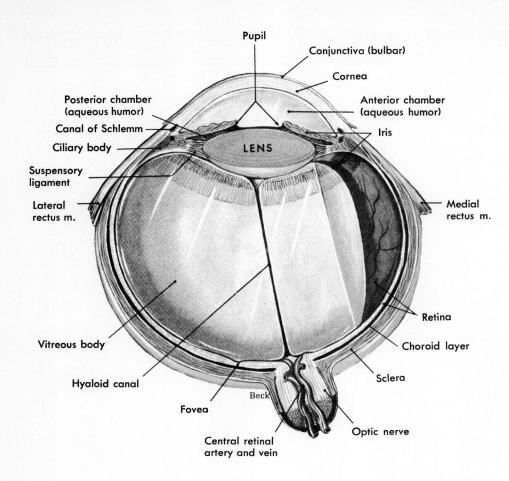

Pupil

Conjunctiva (bulbar)

Cornea

Posterior chamber
(aqueous humor)

Anterior chamber
(aqueous humor)

Canal of Schlemm

Iris

Ciliary body

LENS

Suspensory
ligament

Lateral
rectus m.

Medial
rectus m.

Vitreous body

Retina

Choroid layer

Hyaloid canal

Sclera

Beck

Fovea

Optic nerve

Central retinal
artery and vein

FIG. 50-2 Horizontal section through the left eyeball.

PROCEDURE A   ACCESSORY STRUCTURE OF EYE

1. Using the dissecting animal, locate and name the accessory structures of the eye.
2. Name the function of each accessory structure.

PROCEDURE B   EYEBALL ANATOMY

1. Remove the eyeball from the dissecting animal and remove any fatty tissue.
2. Locate the optic nerve, sclera, and cornea.
3. Make an incision around the cornea and remove; note the characteristics.
4. Observe the iris and pupil; explain the function of each.
5. Cut the eyeball in half, dividing the anterior and posterior portions.
6. Note the vitreous humor, ora serrata, and retina.
7. Remove the retina and observe the choroid.
8. Note the suspensory ligaments of the lens; remove the lens and look through it, making careful observations.
9. Pull apart the lens and explain your observations.
10. Use the dissecting microscope to make observations of each structure.
11. Explain the function of each structure.

1. Using the following diagram, label the name of the eye structure next to the appropriate blank.

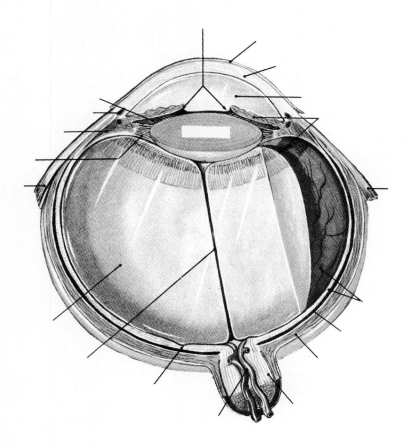

2. Define the following terms:

   Tarsal plate

   Meibomian gland

   Conjunctiva

   Lacrimal gland

*Use all references and materials at your disposal to answer these review questions.

Lysozyme

Ora serrata

Canal of Schlemm

Blind spot

Fovea

3. Explain the following disorders:

Cataract

Glaucoma

Conjunctivitis

Detached retina

Sty

Myopic vision

Hypermetropic vision

Astigmatism

4. Discuss how light rays pass through the cornea, aqueous humor, pupil, lens, and vitreous humor.

5. Explain why the pigment rhodopsin is important to the eye.

51 SPECIAL SENSES
Visual physiology

OBJECTIVES

1 Measure visual acuity using the Snellen chart.
2 Correlate age and the near point of accommodation.
3 Determine an individual's blind spot.
4 Observe binocular vision and convergence.
5 Observe the pupillary reflexes.
6 Observe the afterimage of an object.

MATERIALS

3- x 5-inch index card   small penlight
blind spot diagram card   lens accommodation card
Snellen chart   afterimage card
pencil

The visual formation of an image requires focusing light rays on photoreceptor cells. Light rays entering the eye travel through the cornea, aqueous humor, lens, and vitreous humor. Because of density change of each medium and the curvature of the cornea and lens, the light rays are **refracted,** or bent, so they reach the retinal area where vision is the sharpest, the **central fovea.** The acuteness of vision may be tested using a **Snellen chart.*** On the chart are letters of different sizes. The chart is read at 20 feet (6 meters) away; "20/20 vision" denotes that the subject is 20 feet away and can clearly focus on the line labled "20."

Light rays entering at different angles are **accommodated** by the ability of the lens to curve moderately at one moment and curve greatly at the next moment.

Visual formation of an image also requires the **constricting** of the iris muscles. This prevents light rays from entering the periphery of the lens.

Images forming on the retinas of both eyes must form on the same point of both retinas. This occurs because of the **convergence** of the two eyeballs toward the viewed object as the extrinsic eye muscles coordinate the action. Integration of the images by the brain provides depth perception.

An image may be temporarily seen even though it is no longer viewed. This is called the **afterimage.** It occurs as the retina **photoreceptors** continue to fire off impulses some time after the object is removed.

A well-functioning lens is semispherical. If the lens surface region becomes uneven, there is an unequal bending of light rays, causing **astigmatism.**

---

* Instructor should obtain a Snellen chart from the health center for this procedure.

PROCEDURE A   SNELLEN CHART VISUAL ACUITY (INSTRUCTOR WILL PROVIDE)

1. Stand about 20 feet (6 meters) from the Snellen chart and cover the right eye with an index card.
2. Read down the chart until the letters can no longer be focused.
3. Record the number of the last line that can be read.
4. Repeat the procedure using the left eye.

PROCEDURE B   LENS ACCOMMODATION NEAR POINT

1. Using the lens accommodation card, close one eye and focus on the letter. Measure and record the distance from the eye.
2. Bring the card as close to the open eye as possible so that it is still clear. The distance from the eye to the card is known as the near point of accommodation.
3. Using the chart (Fig. 51-1), compare age and the near point of accommodation.

Correlation of Age and Near Point of Accommodation Near Point		
AGE	CM	INCHES
10	7.5	2.95
20	9.0	3.54
30	11.5	4.53
40	17.2	6.77
50	52.5	20.67
60	83.3	32.80

FIG. 51-1   Lens accommodation near point chart.

PROCEDURE C   BLIND SPOT

1. Hold the blind spot diagram card (Fig. 51-2) about 2 feet (two thirds meter) away from the eyes. Cover the right eye with an index card.
2. While staring at the circle, slowly bring the card toward the face. Note when the cross disappears. This is when the image is on the blind spot of the left eye.
3. Repeat the procedure with the left eye covered.

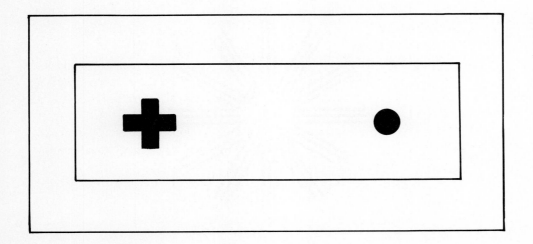

FIG. 51-2   Blind spot test.

PROCEDURE D BINOCULAR VISION AND CONVERGENCE

1. Have your laboratory partner focus on the point of a pencil held about 2 feet (two thirds meter) from his or her nose.
2. Slowly bring the pencil toward your partner's nose until he or she sees two pencil points.
3. Observe the action of your laboratory partner's eye.

PROCEDURE E PUPILLARY REFLEXES

1. Have your laboratory partner cover his or her right eye with an index card.
2. Hold a small penlight about 6 inches (15 cm) from your laboratory partner's left eye. Shine the light into the eye for 2 seconds. Note any changes in the pupil.
3. Wait about 2 minutes and repeat the procedure with the right eye.

PROCEDURE F AFTERIMAGE (Fig. 51-3)

1. Color the afterimage card according to the designated colors.
2. Stare at the cross in the center of the afterimage card for 30 seconds.
3. Immediately stare at an unlined white card. Record your observations. (This may take several tries.)

FIG. 51-3 Afterimage test.

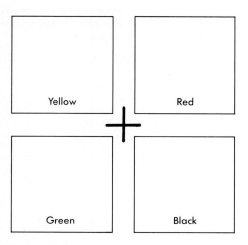

PROCEDURE G ASTIGMATISM

1. Cover one eye and look at the center of the astigmatism chart (Fig. 51-4). Astigmatism is lacking if all the peripheral lines have equal intensity of blackness.
2. Repeat with the other eye.

FIG. 51-4 Astigmatism chart.

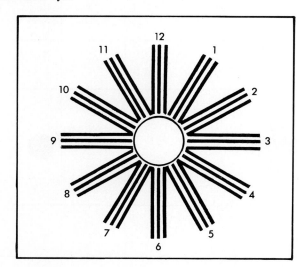

1. Draw and explain the refraction of light rays as they pass through the cornea, aqueous humor, lens, and vitreous humor.

2. Explain the function of the Snellen chart.

3. Why are binocular vision and convergence important?

4. What muscles control the movement of the eyes during convergence?

5. Explain why an individual has a blind spot.

6. Explain what occurs in an afterimage test.

7. Why does the iris smooth muscle change with light intensity?

8. What is astigmatism?

*Use all references and materials at your disposal to answer these review questions.

52

SPECIAL SENSES
Auditory gross anatomy and physiology

OBJECTIVES

1 List the anatomical subdivisions of the ear.
2 Compare the function of the anatomical subdivisions of the ear.
3 Explain the function of the ear related to equilibrium.
4 List the events in the physiology of hearing.

MATERIALS

ear model
dissecting animal

dissecting equipment
vibrating fork
revolving stool

The ear serves as the organ of reception for hearing and equilibrium. It is divided into three anatomical regions: the **external ear,** the **middle ear,** and the **inner ear** (Fig. 52-1). Each anatomical region may be further divided into the following structures:

External ear	Collects sound waves and funnels them inward
Auricle (pinna)	Mainly cartilaginous structure that traps sound waves
External auditory meatus	Lies in external meatus of temporal bone; lined with a few hairs and sebaceous glands, called **ceruminous glands,** which prevent foreign objects from entering ear
Tympanic membrane (eardrum)	Elastic fibrous connective tissue separating external auditory meatus from the middle ear; skin covers external surface and mucous membrane covers internal surface
Middle ear	Hollow cavity in temporal bone
Eustachian tube (auditory tube)	Opening of anterior cavity connecting middle ear to nose and nasopharynx; equalizes air pressure on both sides of tympanic membrane
Malleus (hammer)	Auditory ossicle attached to tympanic membrane
Incus (anvil)	Auditory ossicle that articulates with malleus
Stapes (stirrup)	Auditory ossicle that articulates with incus; fits into small membranous opening between middle and inner ear called **fenestrated vestibuli** (oval window); below oval window is another membranous opening between middle and inner ear called **fenestrated cochlea** (round window)

Inner ear (labyrinth)

Bony (osseous) labyrinth	Series of fluid filled (perilymph) cavities in petrous portion of temporal bone
Vestibule	Separated from middle ear by oval and round windows; central portion of bony labyrinth
Semicircular canal	Superior and posterior of vestibule; three canals at right angles to each other; meet with vestibule at structure called **ampulla**
Cochlea	Bony spiral canal; medial to vestibule; divided into the scala vestibuli and scala tempani with the cochlear duct between the two
Membranous labyrinth	Lines osseous labyrinth and is filled with **endolymph;** space between osseous and membranous labyrinth contains **perilymph**
Semicircular ducts	Contain receptors for dynamic equilibrium and communicate with **utricle** and **saccule**
Utricle	Membrane sac lining of vestibule; communicates with membranous labyrinth of semicircular canals; contains receptors for static equilibrium
Saccule	Membrane sac lining of vestibule; continuous with membranous labyrinth of cochlea; contains receptors for static equilibrium
Cochlear duct	Membranous tubelike extension within the cochlea; begins at junction with oval window
Scala vestibuli	Second channel of cochlea; begins at junction with oval window, filled with perilymph and communicates with the cochlear duct via the vestibular membrane
Scala tympani	Third channel of cochlea; continuous with scala and ends with junction of round window; vestibuli filled with perilymph and communicates with cochlear duct via **basilar membrane,** which contains specialized epithelial cells called **organ of Corti**

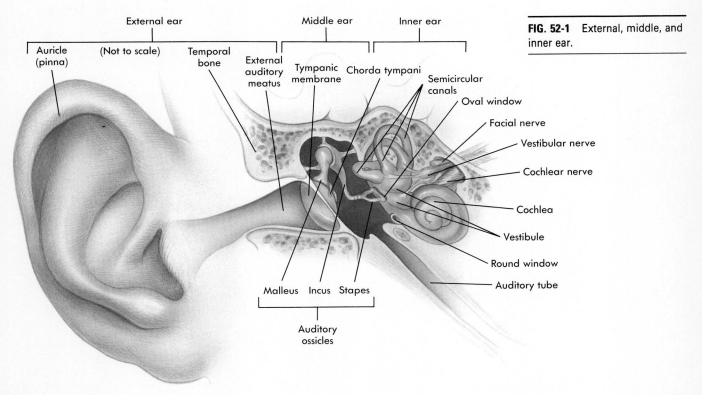

FIG. 52-1 External, middle, and inner ear.

The **cochlear branch** (see Fig. 52-1) of the **vestibulocochlear nerve** is stimulated by impulses from the cochlea for sound wave interpretation. The **vestibular branch** (see Fig. 52-1) of the vestibulocochlear nerve is stimulated by impulses from the vestibule for maintaining equilibrium.

Deafness is associated with poor bone conduction or damage to the organ of Corti or the cochlear nerve. **Nerve deafness** is usually permanent; however, **conduction deafness** may be aided by a hearing aid. To test for nerve or conduction deafness, a vibrating tuning fork is placed in front of the ear. When it can no longer be heard, the tuning fork is placed on the base of the mastoid process. If the sound is heard, the deafness is probably caused by poor bone conduction; however, if the sound cannot be heard, the deafness is probably associated with the organ of Corti or the cochlear nerve.

PROCEDURE A   EXTERNAL EAR IDENTIFICATION

1. Using the ear model and the dissecting animal, locate and identify the external ear structures.
2. Explain the function of each structure identified.

PROCEDURE B   MIDDLE AND INNER EAR IDENTIFICATION

1. Using the ear model, locate and identify the structures of the middle and inner ear.
2. Explain the function of each structure identified.

PROCEDURE C   EAR PHYSIOLOGY

1. Using your laboratory partner, place a vibrating fork on the following cranial bones (one at a time): frontal, occipital, parietal, and temporal. Observe.
2. When the sound can barely be heard, place the tuning fork next to the ear. Compare the conduction of sound in air and bone.

PROCEDURE D   EQUILIBRIUM

1. Sit on a stool and revolve the stool for several seconds.
2. Stop the stool and explain the sensation.

1. Define the following disorders:

   Mastoiditis
   Labyrinth disease
   Impacted cerumen
   Otitis media
   Motion sickness
   Meniere's disease

2. Explain the events that occur in the physiology of hearing sound waves.

3. Explain the difference between static equilibrium and dynamic equilibrium.

4. Use the following diagram to place the appropriate term next to the correct line.

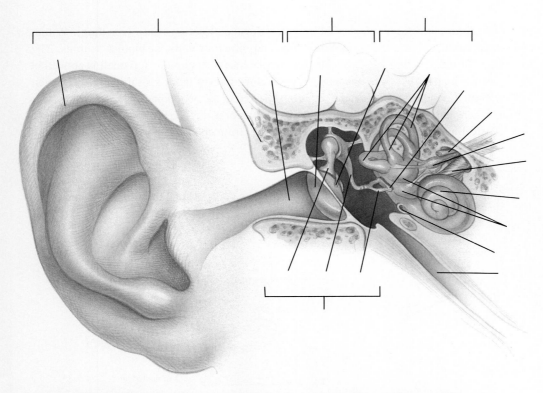

*Use all references and materials at your disposal to answer these review questions.

53 SPECIAL SENSES
Olfactory and gustatory sensations

OBJECTIVES

1 Identify the cellular structures of olfactory sensation.
2 Explain the neural pathway for smell.
3 Determine the location of the four basic taste areas.
4 Describe the cellular structures associated with gustatory receptors.
5 Explain the neural pathway of taste.

MATERIALS

olfactory mucosa prepared slide
tongue prepared slide
compound microscope
dissecting microscope
dissecting animal
dissecting equipment

The nasal epithelium contains the receptors for the **olfactory sense** (smelling). Within the nasal epithelium are the **supporting cells** and **olfactory cells,** or bipolar neurons. The dendrites of the olfactory cells are called **olfactory hairs.** A stimulus for smell is converted to a nerve impulse, which is transferred to **olfactory nerves** that pass through the foramen of the **cribriform plate.** In turn, the impulse is passed to the neural pathway of the **olfactory bulb,** where the impulse is taken to the olfactory area of the cerebral cortex.

Taste buds (Fig. 53-1) located in the tongue are responsible for the main **gustatory**

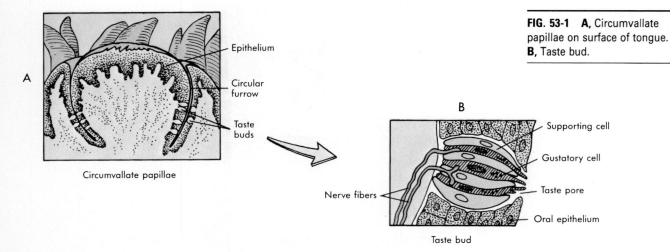

A
Circumvallate papillae

— Epithelium
— Circular furrow
— Taste buds

B

— Supporting cell
— Gustatory cell
— Taste pore
— Oral epithelium

Nerve fibers —

Taste bud

FIG. 53-1 A, Circumvallate papillae on surface of tongue. **B,** Taste bud.

sensations (tasting). The soft palate and tongue also contain some taste buds. Within a taste bud are **supporting cells** and **gustatory cells.** The gustatory cells have a **gustatory hair,** which extends through an opening of the taste bud called the **taste pore.**

Connective tissue elevations of the tongue called **papillae** (see Fig. 53-1) contain the taste buds. Circular papillae located at the posterior portion of the tongue are called **circumvallate papillae.** Mushroom-shaped papillae located on the tip and sides of the tongue are called **fungiform papillae.** Many "tastes" are actually odors; however, there are basically four tastes: sweet, sour, bitter, and salty. **Bitter** substances are easily detected by the posterior portion of the tongue; **sour** substances are easily detected by the lateral portion of the tongue; **sweet** and **salty** substances are easily detected by the tip of the tongue.

Taste impulses are passed along the taste buds to the **facial nerve, glossopharyngeal nerve,** and **vagus nerve.** Impulses are then evaluated by the gustatory area of the cerebral cortex.

PROCEDURE A   OLFACTORY HISTOLOGY

1. Observe an olfactory mucosa prepared slide under high power of the microscope.
2. Draw and label the olfactory epithelium, olfactory cells, supporting cells, and olfactory hairs.
3. Explain the neural pathway for smell.

PROCEDURE B   OLFACTORY SENSATION

1. Place a glass elbow in a one-hole rubber stopper. Attach a rubber hose to the glass elbow.
2. Add a diluted perfume solution to a 250 ml Erlenmeyer flask and plug it with the stopper.
3. Close one nostril and, with the rubber hose about 3 mm (⅛ inch) away from the other nostril, inhale the perfume.
4. Wait 1 minute and repeat the procedure with the rubber hose about 6 mm (½ inch) away from the nostril. Note any differences.
5. Using a vial of oil of cloves, inhale several times.
6. Inhale the perfume solution again. Note any differences.

PROCEDURE C   DISSECTING MICROSCOPIC OBSERVATIONS OF THE TONGUE

1. Remove the dissecting animal tongue and make observations under the dissecting microscope.
2. Compare the four basic taste areas of the tongue.

PROCEDURE D   COMPOUND MICROSCOPIC OBSERVATIONS OF THE TONGUE

1. Observe a tongue prepared slide under high power.
2. Identify papillary structures and taste buds.

PROCEDURE E   GUSTATORY SENSATION

1. Place a drop of sugar solution on the tip of the tongue. Note the taste.
2. Rinse the mouth with water and repeat the procedure on the side of the tongue. Note the taste.
3. Rinse again with water and repeat the procedure on the back of the tongue. Note the taste. Record the region of the tongue in which the taste of sugar was strongest.
4. Repeat these procedures using a salt solution, an acid solution, and a bitter solution.
5. Record the region of the tongue in which the taste of each solution was strongest.

PROCEDURE F   TASTE TEST

1. Place a piece of apple in your mouth and note the taste.
2. Repeat the procedure with your nostrils closed. Note the taste.
3. Using a piece of raw potato, onion, cheese, and bologna repeat these procedures and note the taste of each food.

1. Discuss the various theories that explain the mechanism by which the stimulus for smell is converted.

2. Discuss the neural pathway for smell.

3. Explain the olfactory adaptation.

4. Compare the structure and function of the following papillae:

   Circumvallate papillae

   Fungiform papillae

   Filiform papillae

5. Explain the neural pathway for taste.

6. Identify the tongue regions stimulated by sweet, sour, bitter, and salty solutions.

*Use all references and materials at your disposal to answer these review questions.

ENDOCRINE SYSTEM
Gross anatomy

MATERIALS

human model

The **endocrine system** releases **hormones** into the bloodstream to regulate bodily functions. The nervous system may stimulate or inhibit the release of hormones; likewise, hormones may stimulate or inhibit nerve impulses.

The endocrine glands (Fig. 54-1) examined in this investigation include the following:

Pituitary Referred to as master gland; located in sella turcica of sphenoid bone; connected by **infundibulum** to **hypothalamus;** divided into two lobes—anterior, or **adenohypophysis,** and posterior, or **neurohypophysis;** release or inhibition of hormones from adenohypophysis is controlled by **releasing factors** from hypothalamus; neurohypophysis consists of axons whose cell bodies are located in hypothalamus; hormones produced in hypothalamus are stored in axon terminals of neurohypophysis and released on stimulation

Pineal Located in roof of third ventricle; begins to degenerate about 7 years of age; undetermined physiology

Thyroid Attached to lateral lobes of trachea by mass of tissue called **isthmus;** composed of thyroid follicles containing **principal cells,** which release **thyroxin (T_4)** and **triiodothyronine (T_3)** into the follicular lumen, where it is stored as **thyroid colloid.** Also contains **parafollicular cells** that secrete (thyro)calcitonin.

Parathyroid Two small masses located on each posterior lobe of thyroid gland; consists of **principal cells,** which synthesize **parathyroid hormone**

Thymus Bilobed gland located between lungs and posterior to sternum; divided into lobules, each consisting of cortex and medulla; cortex contains tightly packed lymphocytes; medulla contains scattered lymphocytes; lymphocytes known as **thymus cells (T-cells)** are important to immune system; after puberty, thymus is replaced by fatty tisse

Adrenal Superior to kidney; divided into outer **adrenal cortex** and inner **adrenal medulla;** adrenal cortex contains three zones with each zone secreting different hormones; adrenal medulla consists of **chromaffin cells,** which release hormones that stimulate sympathetic nervous system

Pancreas Posterior and inferior to stomach; **alpha cells** and **beta cells,** found in clusters called **islets of Langerhans,** secrete **glucagon** and **insulin** to regulate blood sugar level

KEY TERMS

Adenohypophysis
Adrenal cortex
Adrenal gland
Adrenal medulla
Alpha cells
Beta cells
Chromaffin cells
Endocrine system
Glucagon
Hormones
Hypothalamus
Infundibulum
Insulin
Islets of Langerhans
Isthmus
Neurohypophysis
Pancreas
Parathyroid gland
Parathyroid hormone
Pineal gland
Pituitary gland
Principal cells
Releasing factors
Thymus cells
Thymus gland
Thyroid colloid
Thyroid gland
Thyroxin
Triiodothyronine

FIG. 54-1 Endocrine glands.

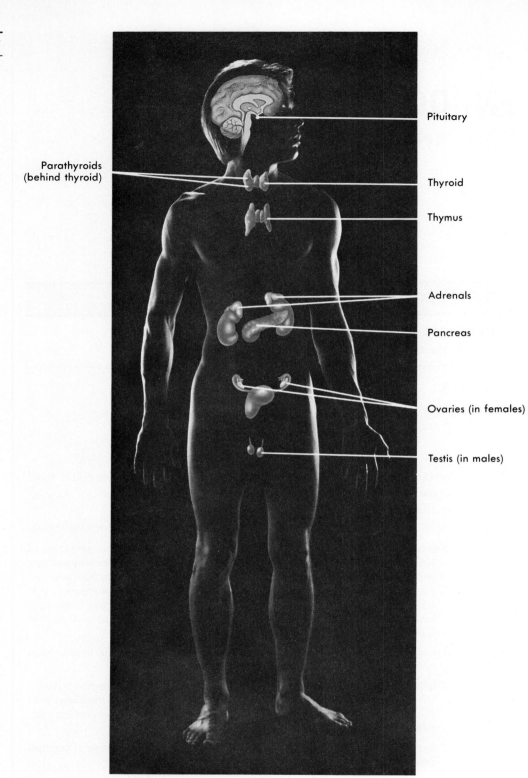

Pituitary

Parathyroids
(behind thyroid)

Thyroid

Thymus

Adrenals

Pancreas

Ovaries (in females)

Testis (in males)

PROCEDURE A   GROSS ANATOMY IDENTIFICATION

1. Locate each endocrine gland.
2. Draw and label the external anatomy of each endocrine gland investigated.

1. Discuss the proposed role of cyclic AMP and the action of genes in hormonal action.

2. Explain the disorders of hyposecretion and hypersecretion. Give a clinical example of each.

3. Define the following terms:

   Dwarfism

   Gigantism

   Acromegaly

4. Compare interactions of the nervous system and the endocrine system.

5. Explain the difference between an endocrine and an exocrine gland.

6. What is the difference between a water-soluble hormone and a lipid-soluble hormone?

*Use all references and materials at your disposal to answer these review questions.

55 ENDOCRINE SYSTEM
Histology

KEY TERMS

Acidophils
Adrenal cortex
Adrenal gland
Adrenal medulla
Alpha cells
Basophils
Beta cells
Chromaffin cells
Endocrine system
Glucagon
Human growth hormone
Insulin
Islets of Langerhans
Pancreas
Pituitary gland
Principal cells
Sinusoid capillaries
Thyroid colloid
Thyroid follicle
Thyroid gland
Thyroxin
Triiodothyronine
Zona fasciculata
Zona glomerulosa
Zona reticularis

OBJECTIVES

1 Identify, draw, and label histological structures.
2 Explain the production and function of hormones of the endocrine system.

MATERIALS

histology slides   compound microscope

The **endocrine system** releases hormones, which are chemical messengers or regulators to control bodily functions. Hormones are usually produced by specialized cells within the endocrine gland. These cells can be identified through histological cross-sections along with special staining techniques. **Acidophils** are the "red" cells. They have a red stained cytoplasm and a purple stained nucleus. **Basophils** are the "blue" cells. They have a blue stained cytoplasm and a purple stained nucleus.

The histological structures are as follows:

Pituitary gland (Color Plate 12, A and B)
Acidophils secrete **human growth hormone (HGH)**; basophils secrete hormones necessary for stimulation of reproductive system and metabolic processes; **sinusoid capillaries** may be identified by small circular "red dots" that are red blood cells within capillaries

Thyroid gland (Color Plate 12, C)
Under high power appears as many circular light pink areas surrounded by cuboidal cells; each circle is known as a **thyroid follicle**; cuboidal cells around lumen are called **principal cells** and manufacture **thyroxin (T_4)** and **triiodothyronin (T_3)** into follicular lumen, where they are stored as **thyroid colloid** (light pink area)

Adrenal gland (Color Plate 12, D)
Contains two regions; outer region, **adrenal cortex,** is divided into three zones: zona glomerulosa, zona fasciculata, and zona reticularis (cells of which secrete steroid hormones); **adrenal medulla** consists of **chromaffin cells,** which release hormones that stimulate sympathetic nervous sytem

Pancreas (Color Plate 12, E)
Cluster area, which appears to be lighter than other components, is known as **islets of Langerhans**; highly granulated cells that stain red-purple are **alpha cells,** which secrete **glucagon;** very lightly stained area around these cells contains **beta cells,** which secrete **insulin**

PROCEDURE A   HISTOLOGY

1. Identify, draw, and label the histological structures of each endocrine slide.

1. What hormones are produced by the pituitary gland?

2. Define the following terms:

   Exophthalmic goiter

   Simple goiter

   Myxedema

   Cretinism

3. Name the gland and give the function of the following hormones:

   Human growth hormone

   Adenocorticotropic hormone

   Follicle-stimulating hormone

   Prolactin

   Melanocyte-stimulating hormone

   Oxytocin

   Antidiuretic hormone

   Calcitonin

*Use all references and materials at your disposal to answer these review questions.

56 CARDIOVASCULAR SYSTEM
Blood histology

KEY TERMS

Agranulocytes
Antibody
Antigen
Antigen-antibody complexes
Antihistamines
Basophils
Blood
Differential count
Eosinophils
Erythrocytes
Formed elements
Globin
Granulocytes
Hematopoiesis
Hemoglobin
Histamine
Interstitial fluid
Iron
Leukocytes
Lymph nodes
Lymphocytes
Megakaryocytes
Monocytes
Neutrophils
Plasma
Plasma cells
Red bone marrow
Spleen
Thrombocytes
Tonsils

OBJECTIVES

1 List the structural features and functions of erythrocytes.
2 Compare the structure and function of leukocytes.
3 Explain the formation and structure of thrombocytes.

MATERIALS

prepared blood smear slides immersion oil
compound microscope

The function of **blood** is to transport vital substances to all tissues. This is done by diffusion of these substances from the capillaries into the **interstitial fluid,** and from there into the cells. Wastes are diffused by the reverse process and carried to areas from which they are eliminated from the body.

Blood is composed of **plasma** and **formed elements,** or blood cells. Plasma is about 90% water and about 10% proteins, nonproteins, food substances, regulatory substances, respiratory gases, and electrolytes.

All blood cells are formed by the process called **hematopoiesis.** In adults, **red blood cells, granulocytes,** and **platelets** are produced in the **red bone marrow. Agranulocytes** are produced in the **spleen, tonsils, lymph nodes,** and **red bone marrow.**

Blood cells may be classified as follows:

**Erythrocytes
(red blood cells)**
(Color Plates 13 and 14)

Biconcave disks, which provide greater surface area; they lack nuclei, and each cell contains about 280 million molecules of **hemoglobin;** oxygen molecules bind with **iron** atom of hemoglobin, and carbon dioxide molecules combine with **globin** proteins of the hemoglobin. Red color of blood results from hemoglobin

Classified further into the following:

**Leukocytes
(white blood cells)**
(see Color Plates 13 and 14)

Granulocytes — Have granular cytoplasm and lobed nucleus

Neutrophils — Contain a nucleus with two to three lobes or segments; small brownish cytoplasmic granules; phagocytic cells that are most active during microbial invasion

Eosinophils — The nucleus usually has two oval lobes; large cytoplasmic granules stain reddish orange; may leave capillaries, where they produce **antihistamines**; high counts indicate allergic conditions as well as **antigen-**

214

	antibody complexes (most **antigens** are proteins foreign to body; **antibodies** are proteins that make antigens nonfunctional)
Basophils	Cytoplasm contains large darkly stained granules that release heparin, an anticoagulant, which prevents intravascular coagulation, and histamine, which increases inflammation
Agranulocytes	Under microscope, cytoplasmic granules are not visible and the nucleus appears spherical
Monocytes	Kidney-shaped nucleus; largest of leukocytes; phagocyte-like neutrophils that take longer to reach site of infection; come in greater numbers to destroy more microbes; increased numbers indicate chronic infection; these mature into macrophages
Lymphocytes	Large spherical nucleus; usually stains blue; these are immunocytes; there are 2 kinds of lymphocytes: B cells and T cells; antigens stimulate B cells to differentiate into antibody-secreting **plasma cells**
Thrombocytes (platelets)	Small, disk-shaped cell fragments that lack nucleus; they arise from pinched off portions of **megakaryocytes;** function to initiate blood clotting

The percentage excess or deficiency of leukocytes may indicate a homeostatic imbalance. A **differential (WBC) count** may be used as an important diagnostic tool, since elevated or depressed levels of specific WBC are often indicative of certain diseases. This is done by counting a total of 100 WBC while classifying them as to type. Moving the slide in a zigzag pattern across the stage prevents counting the same cell twice. A normal differential count may be as follows:

Neutrophils	60% to 70%
Eosinophils	2% to 4%
Basophils	0.5% to 1.0%
Lymphocytes	20% to 25%
Monocytes	3% to 8%
TOTAL	100%

PROCEDURE A   ERYTHROCYTE IDENTIFICATION

1. Obtain a prepared blood smear slide and focus on low power.
2. Locate an erythrocyte; change to oil immersion; draw the structure of the erythrocyte.

PROCEDURE B   LEUKOCYTE IDENTIFICATION

1. Use the same slide to identify any leukocytes as neutrophils, eosinophils, basophils, monocytes, or lymphocytes.
2. Draw and label each type of leukocyte.

PROCEDURE C   THROMBOCYTE IDENTIFICATION

1. Once again, use the slide to identify any platelets.
2. Draw and label the platelets.

PROCEDURE D   DIFFERENTIAL COUNT

1. Use the same prepared blood smear slide to perform a differential count.
2. Record the different kinds of leukocytes up to 100 and give the percentage.

1. Describe the structure of an erythrocyte as it relates to the function.

2. Discuss the structure and function of hemoglobin.

3. Why does an erythrocyte become nonfunctional after about 120 days?

4. Where does the production of erythrocytes occur?

5. Where are erythrocytes destroyed after they are no longer functional?

6. Compare the structure of granulocytes.

7. Compare the function of granulocytes.

8. Explain the different structures of agranulocytes.

9. What are the different functions of agranulocytes?

10. Discuss the structure and function of a thrombocyte.

11. Explain how sickle cell anemia affects the structure and function of a red blood cell.

*Use all references and materials at your disposal to answer these review questions.

CARDIOVASCULAR SYSTEM
Red blood cell homeostasis

57

OBJECTIVES

1 Observe red blood cells in varying osmotic conditions.
2 Explain the rate of hemolysis of red blood cells in various hypotonic solutions.

MATERIALS

screened animal blood
5 clean test tubes
test tube rack
white paper with typing

high-intensity light
0.9% saline solution
distilled water

A red blood cell maintained in optimal environmental conditions is in an **isotonic solution.** Under these conditions, the ratio of solute molecules to solvent molecules is the same on both sides of the red blood cell membrane. When a red blood cell is placed in an environment in which there is a higher concentration of solutes and a lower concentration of water, it is in a **hypertonic solution.** Under these conditions, water molecules inside the red blood cell move out faster than they can enter, causing the cell to shrink. This is called **crenation.** On the other hand, when a red blood cell is placed in an environment in which there is a lower concentration of solutes and higher concentration of water, it is in a **hypotonic solution.*** The water molecules move into the cell faster than they can leave, causing the cell to swell and eventually lyse. This is called **hemolysis.**

Blood and blood solutions have a turbid appearance. In a hypotonic solution, blood cells lyse and fall from the suspension, making the solution clear. The rate at which hemolysis occurs in various hypotonic solutions may be easily determined by placing a printed paper behind a test tube. Because a blood solution is turbid, the letters will appear fuzzy. As hemolysis occurs, the cells will fall from the suspension and the letters may be easily read.

*(Refer to Chapter 6 for a review of isotonic, hypertonic, and hypotonic solutions.)

PROCEDURE A   RED BLOOD CELL HOMEOSTASIS

1. Tape a piece of white paper with typing on the back of a test tube rack.
2. Set up a high-intensity light so that it shines on the test tube rack.
3. Label five clean test tubes #1 to #5 and place the following solution in the appropriate test tube (each tube should end up with a total of 5 ml):

   Tube 1   5.0 ml distilled water

   Tube 2   4.5 ml distilled water
   0.5 ml 0.9% saline solution

   Tube 3   4.0 ml distilled water
   1.0 ml 0.9% saline solution

   Tube 4   3.0 ml distilled water
   2.0 ml 0.9% saline solution

   Tube 5   5.0 ml saline solution

4. Add three drops of blood to test tube #1. Immediately shake and place the tube in the rack. Using a stopwatch, record the time needed for hemolysis to occur. The solution will appear cloudy, then blurred, then clear as hemolysis occurs.
5. Continue with test tubes #2 to #5 and record each time.

REVIEW QUESTIONS*

1. Explain the optimal environmental conditions for a red blood cell.

2. Discuss the effects of a hypertonic solution on a red blood cell.

3. Draw and label the effects of a hypotonic solution on a red blood cell.

4. Compare crenation and hemolysis.

5. In the blood vascular system, what is the salt (NaCl) content of an isotonic solution for red blood cells?

6. In which test tube did hemolysis occur the fastest? Why?

*Use all references and materials at your disposal to answer these review questions.

CARDIOVASCULAR SYSTEM
Heart anatomy

58

KEY TERMS

Aorta
Aortic semilunar valve
Apex
Atrioventricular valve
Bicuspid valve
Chordae tendineae
Coronary sinus
Diastole
Endocardium
Epicardium
Fibrous pericardium
Heart
Inferior vena cava
Interventricular septum
Left atrium
Left ventricle
Mediastinum
Myocardium
Pulmonary semilunar valve
Parietal pericardium
Papillary muscles
Pericardial cavity
Pericardial fluid
Pericardial sac
Pulmonary arteries
Pulmonary veins
Right atrium
Right ventricle
Serous pericardium
Superior vena cava
Systole phase
Tricuspid valve
Visceral pericardium

MATERIALS

dissecting animal   dissecting microscope
dissecting equipment

The **heart** acts as a pump to circulate blood throughout the body. It is located between the lungs in the the **mediastinum**. About two thirds of the mass is to the left of the midline and the other third is to the right. The superior portion of the heart is blunt, whereas the inferior portion is pointed. This portion is called the **apex** and points toward the left of the body.

The heart is protected by two layers of connective tissue called the **pericardial sac,** or **parietal pericardium.** The outer layer, or **fibrous pericardium,** is a loose-fitting connective tissue that prevents overdistension and attaches the heart to the mediastinum. The inner layer is called the **serous pericardium.** Lying directly on top of the heart is a thin layer of connective tissue called the **visceral pericardium,** or **epicardium** (Fig. 58-1). Between the serous part of the parietal pericardium and visceral pericardium is the **pericardial cavity** (see Fig. 58-1) containing **pericardial fluid.** As the heart beats, this fluid prevents friction between the two membranes. The cardiac muscle tissue layer of the heart is called the **myocardium** (see Fig. 58-1). It provides contraction of the heart. The inner layer of the heart, the **endocardium,** (see Fig. 58-1) has two layers. The layer adjacent to the myocardium is made up of connective and smooth muscle tissue. The layer adjacent to the cavity (lumen) of the heart is endothelial tissue.

The interior of the heart is divided into four chambers. The two upper chambers are called the **right atrium** and **left atrium** (Fig. 58-2). The two lower chambers are the **right ventricle** and **left ventricle** (see Fig. 58-2). The right and left chambers are separated by the **interatrial** and **interventricular septum** (see Fig. 58-2). The right atrium receives deoxygenated blood from the body, and the left atrium receives blood from the lungs. The right ventricle pumps deoxygenated blood to the lungs, and the left ventricle pumps oxygenated blood to the body. As the heart contracts, blood leaving the atria or ventricles is prevented from flowing back into the chambers by **valves.**

FIG. 58-1 Histology of heart.

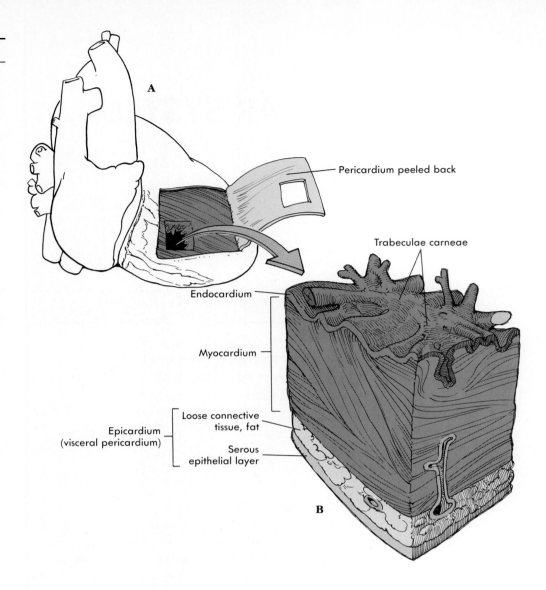

A

Pericardium peeled back

Trabeculae carneae

Endocardium

Myocardium

Epicardium
(visceral pericardium)

Loose connective
tissue, fat

Serous
epithelial layer

B

Valves between the atria and ventricles are known as the **atrioventricular (AV) valves** (see Fig. 58-2). Between the right chambers, the AV valve has three flaps and is called the **tricuspid valve** (see Fig. 58-2). Two flaps form the left AV valve, called the **bicuspid valve** (see Fig. 58-2). The valves extend from the heart chambers and are composed of dense connective tissue covered with endothelium. AV valves are connected to ventricular **papillary muscles,** (see Fig. 58-2) by **chordae tendineae** (see Fig. 58-2), which anchor the valves' cusps to prevent them from bulging up into the atria.

Blood leaving the ventricles entering the great arteries is prevented from flowing back into the heart by **semilunar valves.** The right valve is called the **pulmonary semilunar valve** (see Fig. 58-2), and the left valve is the **aortic semilunar valve** (see Fig. 58-2).

A normal cardiac cycle, or heartbeat, consists of contraction (**systole phase**) and relaxation (**diastole**) of both atria, along with contraction and relaxation of both ventricles. The atria contract simultaneously while the ventricles relax. Then when the ventricles contract, the atria relax. During atrial diastole, blood enters the right atrium from the **superior vena cava, inferior vena cava,** and **coronary sinus,** while blood enters the left atrium from the four **pulmonary veins** (Fig. 58-3). As the atria contract and the ventricles relax, the atrioventricular valves open and blood enters the ventricles. During ventricular contraction, blood leaves the right ventricle through the pulmonary semilunar valve and enters the pulmonary arteries (see Fig. 58-3). Likewise, blood leaves the left ventricle (through the aortic semilunar valve) and enters the **aorta** (see Fig. 58-3).

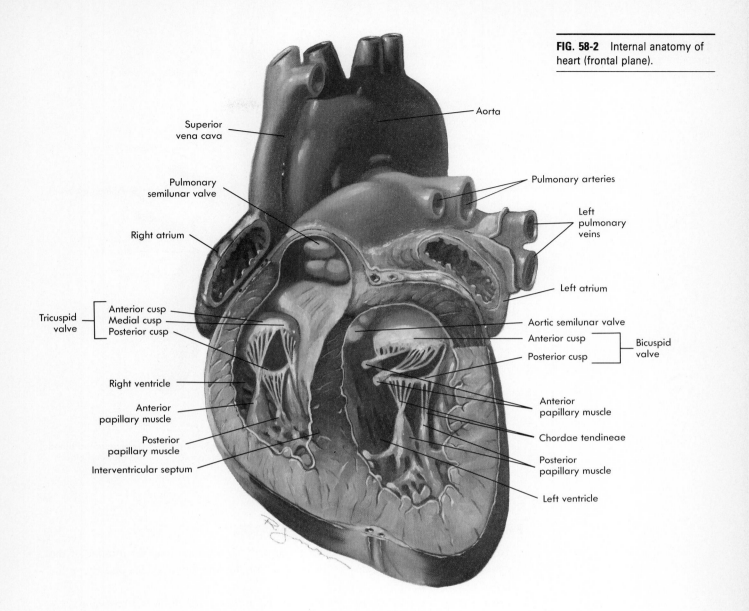

FIG. 58-2 Internal anatomy of heart (frontal plane).

Superior vena cava

Aorta

Pulmonary semilunar valve

Pulmonary arteries

Right atrium

Left pulmonary veins

Left atrium

Tricuspid valve
- Anterior cusp
- Medial cusp
- Posterior cusp

Aortic semilunar valve

Anterior cusp
Posterior cusp
Bicuspid valve

Right ventricle

Anterior papillary muscle

Posterior papillary muscle

Interventricular septum

Anterior papillary muscle

Chordae tendineae

Posterior papillary muscle

Left ventricle

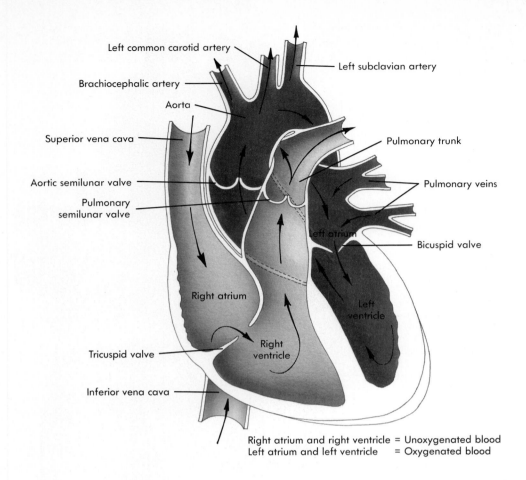

FIG. 58-3 Route of blood flow through heart.

Left common carotid artery

Left subclavian artery

Brachiocephalic artery

Aorta

Pulmonary trunk

Superior vena cava

Pulmonary veins

Aortic semilunar valve

Pulmonary semilunar valve

Left atrium

Bicuspid valve

Right atrium

Left ventricle

Tricuspid valve

Right ventricle

Inferior vena cava

Right atrium and right ventricle = Unoxygenated blood
Left atrium and left ventricle   = Oxygenated blood

PROCEDURE A   EXTERNAL HEART ANATOMY

1. To explore the internal organs, make a longitudinal incision along the midline of the dissecting animal and a tranverse incision below the upper extremities and above the lower extremities.
2. Lift the right and left lobes of the lungs to expose the heart in the mediastinum; describe the size and shape of the heart.
3. Name the external parts of the heart.
4. Identify the arteries and veins entering and leaving the heart.

PROCEDURE B   STRUCTURES OF THE PERICARDIAL SAC AND HEART

1. Remove the intact heart by cutting the blood vessels about 3 cm (1 inch) above and below the heart.
2. Identify the pericardial sac and explain its function; remove the pericardial sac.
3. Make an incision to divide the ventral and dorsal parts of the heart.
4. Using the dissecting microscope, identify and describe the heart wall layers.
5. Name the chambers and valves of the heart.

PROCEDURE C   BLOOD FLOW THROUGH THE HEART

1. Beginning with the atria, demonstrate the flow of blood through the heart.
2. Explain the systole and diastole phases of the chambers during the cardiac cycle.

REVIEW QUESTIONS*

1. Describe the location of the heart.

2. Explain the structure and function of the pericardial sac.

3. Compare the three layers of the heart wall.

4. What is the function of the atria and ventricles?

5. Name and describe the structure and function of the heart valves.

6. Define the following disorders:

   Septal defect

   Valvular stenosis

   Pericarditis

   Heart murmur

7. Explain how an artificial valve works.

8. Explain how pressure in the heart chambers regulate the heart valves.

*Use all references and materials at your disposal to answer these review questions.

59

CARDIOVASCULAR SYSTEM
Cardiac muscle tissue

Endomysium
Epicardium
Intercalated disks
Mesothelium
Muscle fibers
Myocardium
Myofibrils
Sarcolemma
Serous tissue
Striated

OBJECTIVES

1 Compare the structure of the epicardium, myocardium, and endocardium.
2 Identify the microscopic characteristics of cardiac muscle tissue.

MATERIALS

dissecting microscope
compound microscope
dissecting animal
dissecting equipment

cardiac muscle prepared slides
(longitudinal section and
cross-section)

The heart wall consists of three layers. The outer layer, composed of **serous tissue** and **mesothelium,** is called the **epicardium** (see Fig. 58-3). The inner layer, made up of connective tissue, smooth muscle, and blood vessels, is called the **endocardium** (see Fig. 58-1). The middle layer contains cardiac muscle tissue and is called the **myocardium** (see Fig. 58-1). This layer provides rapid and continuous contraction and relaxation of the heart.

In a compound microscope a longitudinal section of **cardiac muscle fibers,** the top and bottom **sarcolemma,** or plasma membrane are undefined. Therefore the fibers appear to branch (Color Plate 15). At the ends of each fiber is an irregular transverse thickening of the sarcolemma known as **intercalated disks** (see Color Plate 16). Cardiac muscle fibers have a single nucleus that is centrally located. The **myofibrils** are organized and provide the same **striated** appearance (see Color Plate 16) as seen in skeletal muscle fibers.

PROCEDURE A   STRUCTURE OF THE HEART WALL

1. Remove the heart from the dissecting animal at the superior and inferior blood vessels.
2. Cut the heart to divide the ventral and dorsal sections.
3. Using the dissecting microscope, identify and describe the three layers of the heart wall.

PROCEDURE B CROSS-SECTION OF CARDIAC MUSCLE TISSUE

1. Obtain a prepared slide of a cardiac muscle tissue cross-section.
2. Observe the muscle tissue on low power and identify the endomysium, sarcolemma, and nucleus.
3. Draw and label the microscopic cross-section of cardiac muscle tissue.

PROCEDURE C LONGITUDINAL SECTION OF MUSCLE TISSUE

1. Obtain a prepared slide of cardiac muscle tissue longitudinal section.
2. Observe the muscle tissue on low power and identify the endomysium, sarcolemma, nucleus, and striations.
3. Draw and label the microscopic longitudinal section of cardiac muscle tissue.

REVIEW QUESTIONS*

1. Explain the function of cardiac muscle tissue.

2. Compare the structure of cardiac muscle tissue with skeletal muscle tissue.

3. Define the disorders:

   Epicarditis

   Myocarditis

   Endocarditis

4. Under normal conditions, how many times does the cardiac muscle tissue contract and relax in 1 minute?

5. How does the source of stimulation for cardiac muscle tissue differ from skeletal muscle tissue?

6. What is the function of the intercalated disks of the cardiac muscle tissue?

*Use all references and materials at your disposal to answer these review questions.

60 SMOOTH MUSCLE HISTOLOGY

KEY TERMS

Actin
Myosin
Smooth muscle tissue

OBJECTIVES

1 Identify smooth muscle tissue microscopically.
2 List the body components associated with smooth muscle tissue.

MATERIALS

dissecting animal
dissecting equipment
dissecting microscope

compound microscope
smooth muscle tissue prepared
  slides

Microscopically, **smooth muscle tissue** (Color Plate 16) appears spindle-shaped with a single, oval, centrally located nucleus. Smooth muscle fibers contain the myofilaments **actin** and **myosin;** however, the filaments are not as organized as in skeletal and cardiac muscle. This characteristic seems to provide smooth muscle tissue with slow contraction and relaxation.

Smooth muscle tissue is located in the walls of blood vessels, iris of the eye, and the walls of most muscular organs such as the stomach, intestine, uterus, and urinary bladder.

PROCEDURE A STEREOSCOPIC OBSERVATION OF SMOOTH MUSCLE TISSUE

1. Using the dissecting animal, remove a section of the stomach and intestine.
2. Make a cross-section and longitudinal section of each tissue.
3. Make observations using the dissecting microscope.
4. While looking in the dissecting microscope, "tease" the smooth muscle tissue.
5. Draw and label observations.

PROCEDURE B COMPOUND MICROSCOPIC OBSERVATION OF SMOOTH MUSCLE TISSUE

1. Obtain a smooth muscle tissue prepared slide and focus on high power.
2. Draw and label observations.

1. In what layer of the lower digestive tract is smooth muscle found?

2. Explain the function of the smooth muscle layer of the lower digestive tract.

3. Describe the smooth muscle fiber arrangement of the iris.

4. Explain the function of the smooth muscle of the iris.

5. In which layer (tunic) of the blood vessels is smooth muscle found?

6. Explain the function of the smooth muscle layer of a blood vessel.

7. Compare the shape, nuclei location, and myofibril arrangement of smooth muscle, cardiac muscle, and skeleton muscle tissue.

8. Compare visceral muscle tissue and multiunit smooth muscle tissue.

*Use all references and materials at your disposal to answer these review questions.

61 CARDIOVASCULAR SYSTEM
Arteries

OBJECTIVES

1 Identify the principal arteries of systemic circulation.
2 Trace the route of the descending aorta, aortic arch, thoracic aorta, and abdominal aorta.

MATERIALS

dissecting animal   dissecting equipment

Systemic circulation is the transport of oxygenated blood from the left ventricle of the heart to all parts of the body except the lungs, and the return of deoxygenated blood to the right atrium of the heart. As blood is pumped out of the left ventricle, it enters the **aorta** and branches into the **ascending aorta, arch of the aorta, thoracic aorta,** and **abdominal aorta.** Each branch of the aorta further branches into major **arteries,** which supply various regions of the body with nutrients and oxygen. The major arteries of the body are listed in Table 61-1 (Figs. 61-1 and 61-2).

PROCEDURE A   ARTERIES OF SYSTEMIC CIRCULATION

1. Using dissecting equipment, locate the aorta and follow the main branches and arteries associated with those branches.
2. Name the body parts associated with each artery.

TABLE 61-1 MAJOR ARTERIES OF THE BODY

ARTERY	REGION SUPPLIED
ASCENDING AORTA	
Coronary arteries	Heart
ARCH OF AORTA	
Subclavian	Upper extremities
Brachial	Brachial
Ulnar	Ulnar
Radial	Radial
Palmar	Palm
Digital	Digits
Vertebral	Vertebrae
Basilar	Posterior brain structures
Common carotids	Head and neck
External carotid	Thyroid gland, tongue, throat, face, ear, scalp, and dura mater
Internal carotid	Brain, eyes, forehead, and nose
THORACIC AORTA	
Intercostals	Intercostal and chest muscles; pleurae
Superior phrenics	Diaphragm
Bronchials	Bronchi of lungs
Esophageals	Esophagus
ABDOMINAL AORTA	
Inferior phrenics	Diaphragm
Common hepatic	Liver
Gastric	Stomach and esophagus
Splenic	Spleen, pancreas, and stomach
Superior mesenteric	Small intestine, cecum, ascending and transverse colons
Suprarenals	Adrenal glands
Renals	Kidneys
GONADALS	
Testiculars	Testes
Ovarians	Ovaries
Inferior mesenteric	Transverse, descending, and sigmoid colons and rectum
Lumbars	Spinal cord and its muscles and meninges
Middle sacral	Sacrum, coccyx, gluteus maximus muscle, and rectum
COMMON ILIACS	
External iliacs	Lower extremities
Internal iliacs	Uterus, prostate, muscles of buttocks, urinary bladder

FIG. 61-1 Cat arteries, ventral and anterior views.

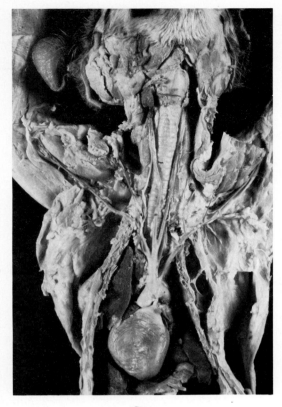

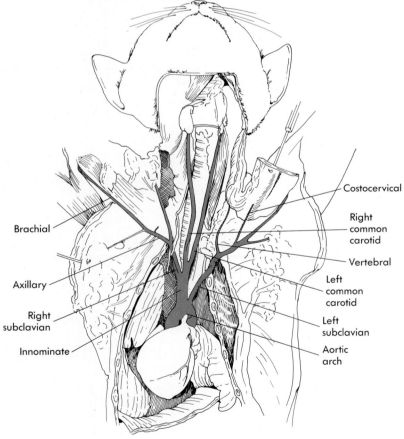

Costocervical

Right
common
carotid

Vertebral

Left
common
carotid

Left
subclavian

Aortic
arch

Brachial

Axillary

Right
subclavian

Innominate

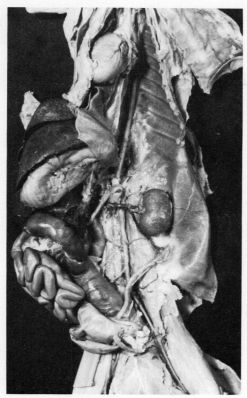

FIG. 61-2 Cat arteries, ventral and posterior views.

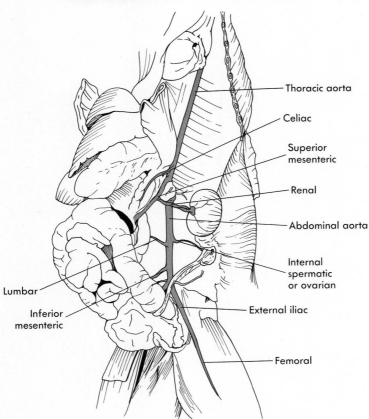

Thoracic aorta

Celiac

Superior mesenteric

Renal

Abdominal aorta

Internal spermatic or ovarian

Lumbar

Inferior mesenteric

External iliac

Femoral

1. Describe the route of blood from the heart to the thyroid gland.

2. Trace the route of blood from the heart to the hand.

3. Trace the route of blood from the heart to the toes.

4. Place the correct artery name next to the corresponding blank.

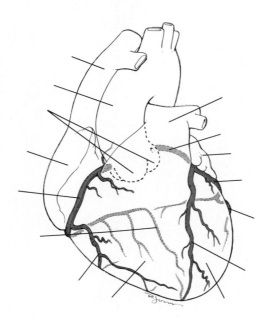

5. Explain the purpose of anastomosis.

*Use all references and materials at your disposal to answer these review questions.

CARDIOVASCULAR SYSTEM
Veins

62

OBJECTIVES

1 Identify the principal veins of systemic circulation.
2 Trace the venous route of the lower extremities, abdomen, pelvis, thorax, upper extremities, neck, and head.
3 Explain the route of the digestive system veins.

MATERIALS

dissecting animal   dissecting equipment

Once blood obtains nutrients and oxygen, it is transported to tissues and returned to the right atrium of the heart through the **systemic veins,** listed in Tables 62-1 through 62-5 (Figs. 62-1 and 62-2).

Blood returns from the head and neck veins through the **subclavian veins** (the subclavian veins unite to form the brachiocephalic veins and the brachiocephalic veins unite to form the superior vena cava, which empties into the **right atrium**) (Table 62-1).

TABLE 62-1   VEINS OF HEAD AND NECK

VEINS	REGION DRAINED
Internal jugulars	Sinuses of the brain
Superior sagittal sinus	Superior part of face and neck
Inferior sagittal sinus	
Straight sinus	
Sigmoid sinus	Brain
External jugulars	Parotid glands, facial muscles, scalp, other superfacial structures
Transverse jugulars	Vertebrae

KEY TERMS

Ascending lumbar vein
Axillary
Azygous
Basilic
Brachials
Brachiocephalic vein
Cephalics
Common iliac
Dorsal arch
Dorsalis pedis
External iliac
External jugular
Femoral
Hepatic
Hepatic portal circulation
Histones
Inferior phrenic
Inferior sagittal sinus
Inferior vena cava
Internal jugular
Lumbar vein
Medial cubital
Ovarian
Peroneal
Popliteal
Renal
Right atrium
Saphenous
Sigmoid sinus
Straight sinus
Subclavian vein
Superior sagittal sinus
Superior vena cava
Systemic veins
Testicular
Tibial
Transverse sinus
Vertebral

233

FIG. 62-1 Cat veins, ventral and anterior views.

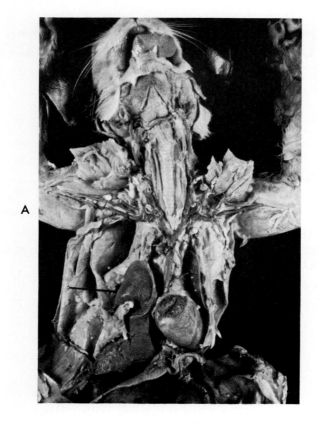

A

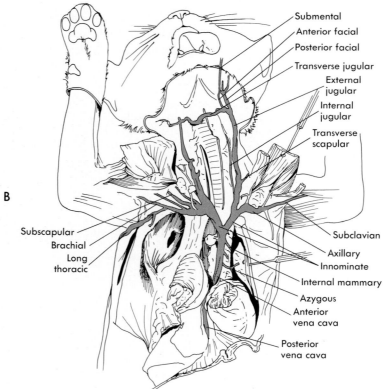

B

Submental
Anterior facial
Posterior facial
Transverse jugular
External jugular
Internal jugular
Transverse scapular

Subscapular
Brachial
Long thoracic

Subclavian
Axillary
Innominate
Internal mammary
Azygous
Anterior vena cava

Posterior vena cava

FIG. 62-2 Cat veins, ventral and posterior views.

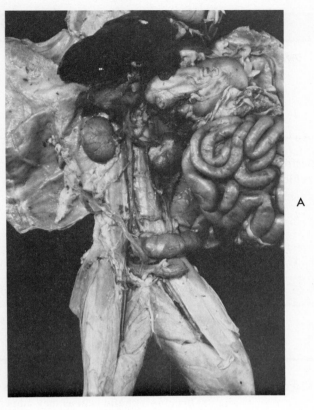

A

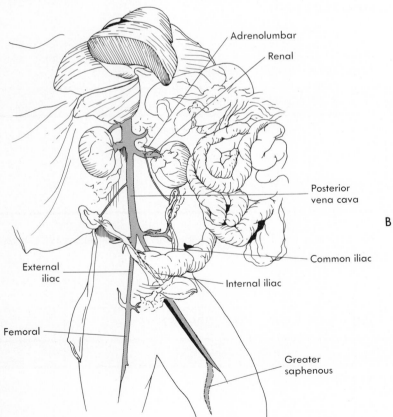

B

Adrenolumbar

Renal

Posterior
vena cava

Common iliac

External
iliac

Internal iliac

Femoral

Greater
saphenous

TABLE 62-2   VEINS OF THE UPPER EXTREMITIES

VEINS	REGION DRAINED
Cephalics	
Dorsal arch	Hand upward around radius
Basilics	Hand upward around ulna
Medial cubital	Used for injection, transfusion, or blood samples
Brachials	
Axillaries	

Blood returns from the upper extremity veins through the **subclavian veins,** which unite to form the **brachiocephalic veins** (the brachiocephalic veins unite to form the superior vena cava, which empties into the **right atrium**) (Table 62-2).

TABLE 62-3   VEINS OF THE THORAX

VEINS	REGIONS DRAINED
Hemiazygous	Left bronchial vein, lumbar vein, intercostal veins, esophageal veins
Azygous	Intercostal veins
	Pericardial veins, right bronchial vein, esophogeal veins

The veins of the thorax continue from the **ascending lumbar veins** and collect blood from the thorax. These veins join the **brachiocephalic vein** as they empty into the **superior vena cava** (Table 62-3).

TABLE 62-4   VEINS OF THE ABDOMEN AND PELVIS

VEINS	REGIONS DRAINED
Lumbar veins	Posterior abdominal wall
Hepatic	Liver
Inferior phrenic	Diaphragm
Ovarian	Ovaries
Testicular	Testes
Renal	Kidneys

The abdominal and pelvic veins join the **inferior vena cava** where blood drains into the **right atrium** (Table 62-4).

TABLE 62-5   VEINS OF THE LOWER EXTREMITIES

VEINS	REGIONS DRAINED
Dorsalis pedis	Foot
Tibial	Tibial
Peroneal	
Popliteal	
Saphenous	
Femoral	Dorsalis pedis, tibial, popliteal, peroneal

The lower extremity veins join the **external iliac,** which in turn joins the **common iliac** and then the **inferior vena cava** (Table 62-5).

The transport of blood from the digestive system is known as **hepatic portal circulation.** This is where deoxygenated, but nutrient rich, blood from the stomach, intestine, pancreas, spleen, and gallbladder is transported to the liver. If necessary, in the liver, substances absorbed from these organs may be stored or detoxified. The blood is then returned to general circulation by way of the hepatic veins emptying into the inferior vena cava.

PROCEDURE A   VEINS OF SYSTEMIC CIRCULATION

1. Using dissecting equipment, locate the major veins.
2. Name the regions drained by the major veins.

PROCEDURE B   VEINS OF HEPATIC PORTAL CIRCULATION

1. Using the dissecting equipment, trace the veins of the hepatic portal circulation to the liver.

REVIEW QUESTIONS*

1. Trace the flow of venous blood from the pancreas to the right atrium of the heart.

2. Explain the function of hepatic portal circulation.

3. Trace the return of blood from the foot.

4. Describe the route of blood from the superior sagittal vein to the superior vena cava.

*Use all references and materials at your disposal to answer these review questions.

5. Place the correct vein name next to the corresponding blank.

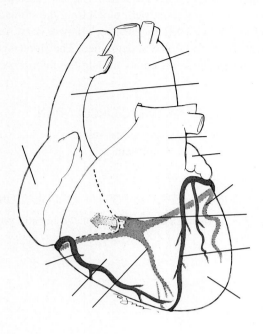

CARDIOVASCULAR SYSTEM
Blood vessel histology

63

OBJECTIVES

1 Compare the structures of arteries, capillaries, and veins.
2 Explain how the structures of arteries, capillaries, and veins relate to their function.

KEY TERMS

Arteries
Arterioles
Blood vessel
Capillaries
Lumen
Tunica externa
Tunica interna
Tunica media
Valves
Veins
Venules

MATERIALS

stereomicroscope
compound microscope
dissecting animal
dissecting equipment

prepared slides of arteries, veins, and capillaries (cross-section and longitudinal section)

Blood vessels provide a transport system in which blood is transported away from the heart to various body tissues and then back to the heart. Vessels that transport blood away from the heart are called **arteries.** Arteries branch off into smaller vessels called **arterioles,** which further branch into **capillaries** as they enter a tissue. Here the exchange of nutrients, wastes, and gases to and from the cells occurs. Capillaries reunite to form small vessels called **venules,** which in turn form larger vessels called **veins.** Veins return blood to the heart.

When blood vessels are compared structurally, arteries are thicker and stronger than veins (Fig. 63-1). This is because arteries have greater blood pressure than veins.

The arterial wall has three layers (Fig. 63-2). The inner layer, **tunica intima,** has sublayers of simple squamous epithelium (endothelium), loose connective tissue, and elastic tissue. The middle layer, **tunica media,** contains elastic and smooth muscle fibers to provide contractility and elasticity as blood flows through the **lumen,** or hollow space, of the blood vessels. The outer layer, **tunica externa,** is composed of loose connective tissue with elastic and collagenous fibers and some smooth muscle tissue.

Veins are composed of basically the same layers as arteries; however, the tunica media is thinner with less connective tissue and smooth muscle. To prevent backflow of blood especially in the limbs, veins contain **valves** (Fig. 63-3). The valves are an extension of the tunica interna and tunica media.

Capillaries are structurally different from arteries and veins. They are an extension of a single layer of endothelium of the tunica interna. The single layer of endothelium allows exchange of nutrients, wastes, and gases.

239

FIG. 63-1 Structural comparison of **A,** artery, and **B,** vein.

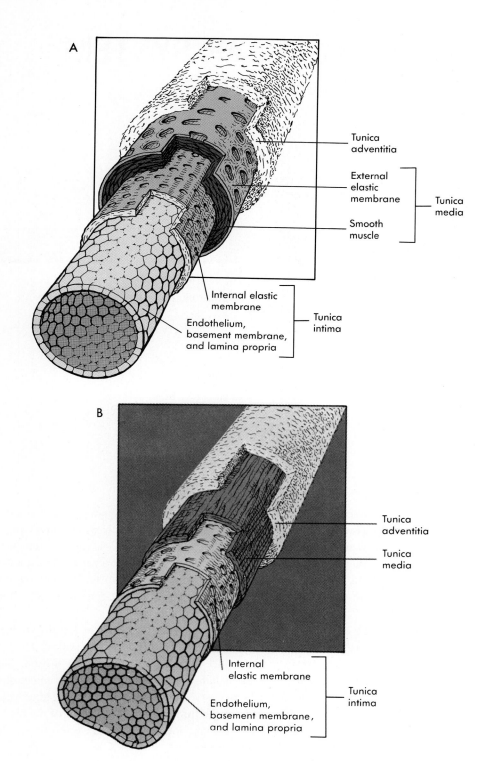

A

Tunica adventitia

External elastic membrane

Tunica media

Smooth muscle

Internal elastic membrane

Tunica intima

Endothelium, basement membrane, and lamina propria

B

Tunica adventitia

Tunica media

Internal elastic membrane

Tunica intima

Endothelium, basement membrane, and lamina propria

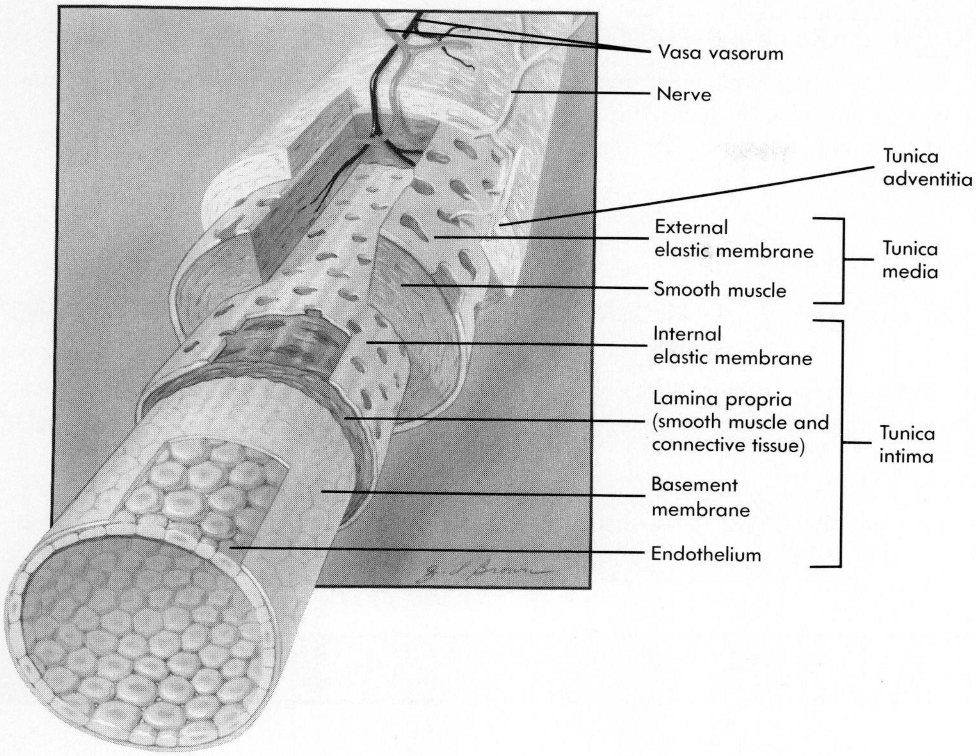

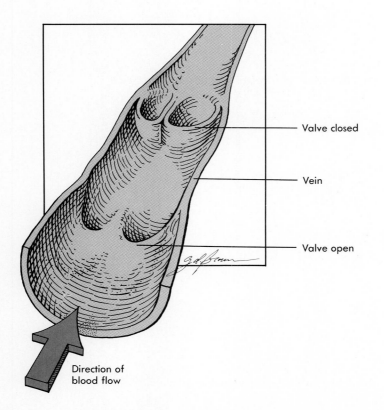

Vasa vasorum

Nerve

FIG. 63-2 Histology of blood vessel.

Tunica adventitia

External elastic membrane ⎤
                      ⎬ Tunica media
Smooth muscle   ⎦

Internal elastic membrane ⎤
                          ⎬
Lamina propria   ⎬ Tunica intima
(smooth muscle and   ⎬
connective tissue)   ⎬
                          ⎬
Basement membrane   ⎬
                          ⎬
Endothelium   ⎦

FIG. 63-3 Valves in veins.

Valve closed

Vein

Valve open

Direction of blood flow

PROCEDURE A   STEREOSCOPIC COMPARISION OF ARTERIES AND VEINS

1. Using dissecting equipment, remove one 10 mm section of an artery and one 10 mm section of a vein (including the valve).
2. Make a longitudinal section; place both sections under the stereomicroscope and observe.
3. Compare the artery and vein.

PROCEDURE B   COMPOUND MICROSCOPIC COMPARISON OF ARTERIES AND VEINS

1. Obtain a prepared slide of the blood vascular system.
2. On high power, make comparisons of the layers of arteries and veins in longitudinal section.
3. Draw and label.
4. On high power, make comparisons of the layers of arteries and veins in cross-section.
5. Draw and label.

REVIEW QUESTIONS*

1. Define the following terms:

   Artery
   Arteriole
   Capillary
   Venule
   Vein
   Tunica interna
   Tunica media
   Tunica externa

2. Explain how the structures of arteries, capillaries, and veins relate to their function.

3. Compare vasoconstriction and vasodilation.

4. Explain the following disorders:

   Artherosclerosis
   Aneurysm
   Varicose veins
   Phlebitis

*Use all references and materials at your disposal to answer these review questions.

CARDIOVASCULAR SYSTEM
Pulse pressure, blood pressure, and electrocardiogram

64

OBJECTIVES

1 Define and explain pulse.
2 Define blood pressure.
3 Explain the procedure for recording blood pressure.
4 Explain how an electrocardiogram records the cardiac cycle.

KEY TERMS

Atrioventricular bundle
Atrioventricular node
Blood pressure
Brachial artery
Diastolic phase
Electrocardiogram
Pacemaker
Pulse
Purkinje fibers
P wave
QRS wave
Sinoatrial node
Sphygmomanometer
Stethoscope
Systole phase
T wave

MATERIALS

laboratory partner   stethoscope
sphygmomanometer   electrocardiograph

Pulse is the alternate expansion and elastic recoil of an artery during contraction and relaxation of the left ventricle. Arteries closest to the heart have the strongest pulse. The pulse measures the rate of the heartbeat. The average resting state pulse rate is between 70 and 90 beats per minute. When a pulse is taken, each beat should be equal in length and strength.

Clinically, **blood pressure** measures the pressure exerted on the arterial walls as the left ventricle contracts **(systole phase)** and the remaining pressure on the arterial walls as the left ventricle relaxes **(diastolic phase)**. Blood pressure is measured by a **sphygmomanometer** (Fig. 64-1).

The cuff of a sphygmomanometer is wrapped around the arm at the left **brachial artery** and inflated by squeezing the bulb. As the cuff is inflated, the artery walls are compressed tightly against each other, preventing blood flow. If a **stethoscope** is placed over the artery below the cuff, no pulse will be detected while the blood flow is obstructed. As air is released from the cuff, the artery will open and recoil. This represents the systole phase of the left ventricle. The sound of recoil may be detected by the stethoscope.

When the sound is heard, a reading on the mercury column is recorded. This is known as the **systolic blood pressure.** As the pressure from the cuff is released, the sound becomes faint, then finally disappears. This measures the force of blood against the arterial wall during ventricular relaxation. A reading on the mercury column is recorded. This is known as the **diastolic blood pressure.** Normal blood pressure is 120 (systolic) over 80 (diastolic).

Contraction of the heart is stimulated by specialized cardiac fibers sending an electrical impulse throughout the heart. The cardiac cycle (Fig. 64-2) is initiated by the **sinoatrial node** or **pacemaker,** located in the superior region of the right atrium. These conducting cells generate a depolarization wave that spreads throughout the two atria. The contractile events follow the electrical events. The impulse is carried to the **atrioventricular node,**

243

located in the inferior region of the right atrium. From here the impulse travels to the **atrioventricular bundle,** located in the intraventricular septum. Finally the impulse reaches the **Purkinje fibers,** which are embedded throughout the myocardium of the ventricles. Here depolarization and contraction of the ventricles occur. At the same time repolarization and relaxation of the atria occur.

Each portion of the cardiac cycle produces a different electrical impulse and may be recorded as waves on an **electrocardiogram (ECG)** (Fig. 64-3). The first upward wave indicates atrial depolarization and is called the **P wave.** The second wave is the **QRS wave** and represents ventricular depolarization. The third wave is the **T wave** and indicates ventricular repolarization. Atrial repolarization is masked by the strong ventricular depolarization.

FIG. 64-1   Blood pressure measurement using sphygmomanometer.

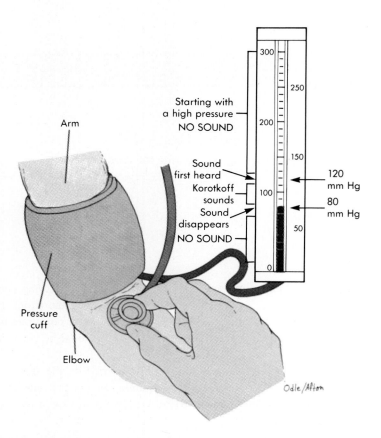

Arm

Starting with a high pressure
NO SOUND

300
250
200
150

Sound first heard
Korotkoff sounds
100
120 mm Hg
80 mm Hg

Sound disappears
NO SOUND
50

Pressure cuff

0

Elbow

Odle/Afton

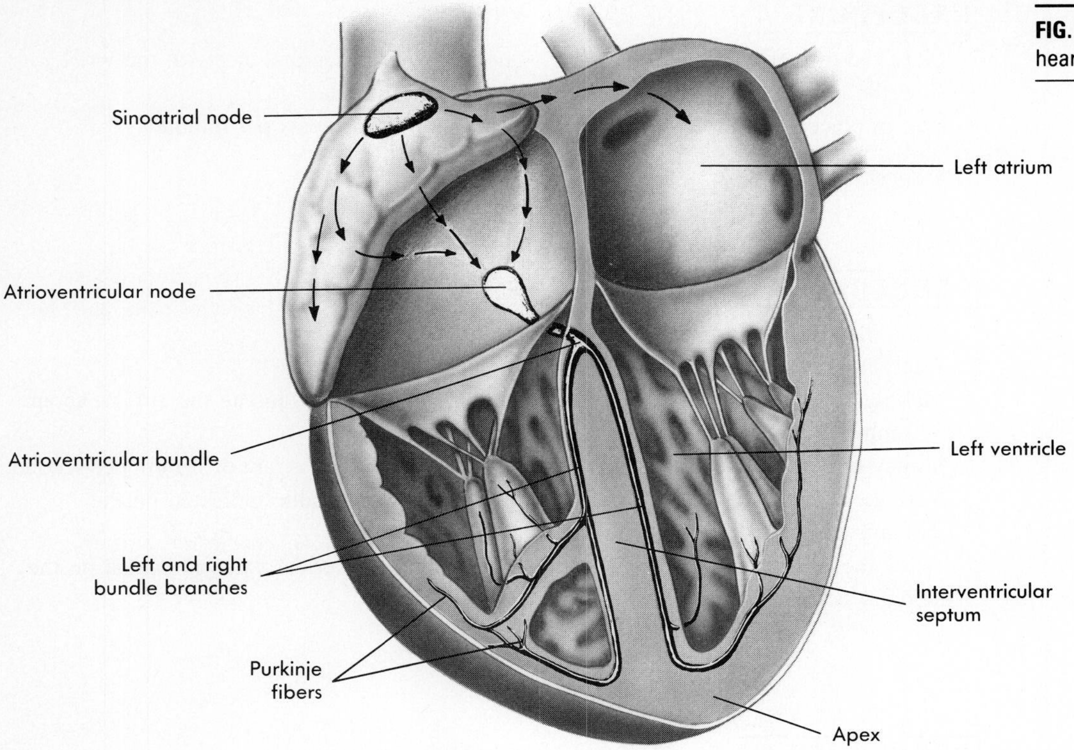

FIG. 64-2 Conducting system of heart.

Sinoatrial node

Left atrium

Atrioventricular node

Atrioventricular bundle

Left ventricle

Left and right bundle branches

Interventricular septum

Purkinje fibers

Apex

FIG. 64-3 Electrocardiogram.

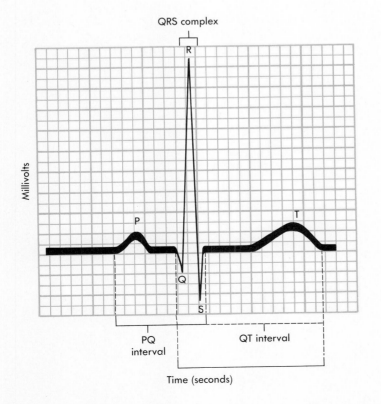

QRS complex

R

P

T

Q

S

Millivolts

PQ interval

QT interval

Time (seconds)

PROCEDURE A   PULSE RATE

1. Place the three middle fingers of the right hand over the radial artery at the wrist.
2. Find the pulse and record the number of beats per minute.
3. Run in place for 2 minutes and compare the number of beats per minute.

PROCEDURE B   BLOOD PRESSURE

1. Place the sphygmomanometer around the left brachial artery.
2. Place the stethoscope on the artery just below the cuff.
3. Making sure the valve of the sphygmomanometer is closed, inflate the cuff to about 200 mm Hg.
4. Slowly open the valve to release pressure from the cuff at a rate of 3 mm/per second. Note the first pulse and sound and record the numerical value indicated on the mercury column for the systolic reading.
5. When the last pulse and sound is heard, record the numerical value indicated on the mercury column for the diastolic reading.

PROCEDURE C   ELECTROCARDIOGRAM

1. Follow proper procedures for the use of the ECG and record the electrical impulses of the cardiac cycle.
2. Use the ECG tape to mark the P wave, QRS wave, and T wave.

1. Define the following terms:

   Tachycardia

   Brachycardia

2. Define systole and diastole. What is their relationship to the cardiac cycle?

3. Describe and explain the function of an artificial pacemaker.

4. Compare atrial flutter, atrial fibrillation, and ventricular fibrillation.

5. Define the following terms:

   Angina pectoris

   Myocardial infarction

6. Draw and label a normal ECG.

7. Explain the diagnostic importance of an ECG.

8. Compare a normal and abnormal pulse.

*Use all references and materials at your disposal to answer these review questions.

65 CARDIOVASCULAR SYSTEM
Blood typing

KEY TERMS

Antibody
Antigen
Lectin
Subgroups
Type A blood
Type B blood
Type O blood

OBJECTIVES

1 **Explain ABO blood typing.**
2 **Discuss antigen-antibody reaction.**

MATERIALS

slide
animal blood types*

lectin*
microspatula

*Courtesy of Plant Something
Different, Inc., P.O. Box
1032, Angleton, TX 77516-
1032.

The word agglutination is used, among others, to refer to the process of red blood cells clumping together to form large masses. This is not formation of a clot. The clumping together of the cells in this manner results from the presence of sugar-protein structures called **antigens** located on the cell surface. Other proteins in the liquid surrounding the cells may have structures that will permit them to lock onto the antigens in very much the same manner that a key fits a lock. These proteins are called **antibodies.** If they attach to the cell proteins, the result will be a linking together of the cells in large clumps. If a person with type A blood received a transfusion from a person with type B, his or her body would consider the B protein a foreign substance, and antibodies to it would cause the B cells to clump together in agglutination.

Agglutination reactions of this kind are commonly used today to determine a person's blood type. In such tests specific antibodies to protein A, protein B, and other blood protein antigens are used. If a person's blood cells agglutinate in the presence of anti-A antibodies, the person has **type A** blood (Fig. 65-1). If they agglutinate with anti-B, the person has **type B** blood. If the cells agglutinate with both anti-A and anti-B, the person has type AB blood. If they agglutinate with neither anti-A nor anti-B, the person is said to have **type O** blood, which means the person has neither A nor B proteins on the cell surface. Although the A and B proteins are by far the most common, many others can be typed in the same way. These are sometimes referred to as the **subgroups.**

In addition to such specific antibodies produced in humans and lower animals, other substances show antibody-like activity. A major group of these is called **lectins.** These are complex large molecules of carbohydrate combined with protein that are produced in some plants and other organisms. One of the characteristics of lectins is that they bring about an agglutination reaction with red blood cells. Additionally, they have many interesting and highly specific attractions for molecules on surfaces of other types of cells. The function of lectins in the plants producing them is relatively unknown. Lectins, like immune system antibodies, are highly specific in their ability to combine various protein and sugar structures.

FIG. 65-1 ABO blood groups.

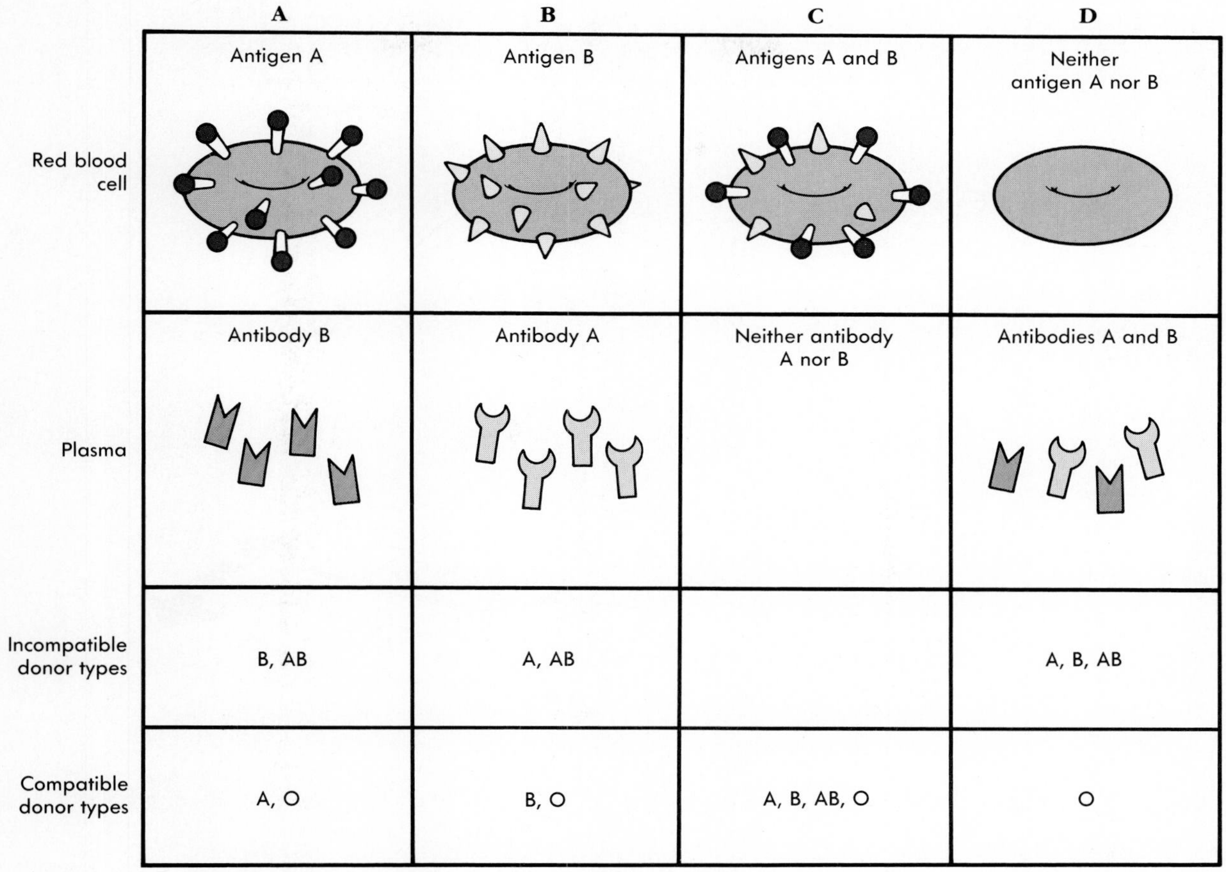

In this experiment the student will take advantage of this specificity to use a plant lectin to identify the species of animal that produced the blood specimens used.

The plant lectin used is reactive with the blood of two of the three animal specimens used. It will cause a very strong agglutination with one of the specimens and more moderate reaction with the other. The third will show no reaction at all, because the red cells do not have on their surface a carbohydrate or protein that will react with the lectin.

PROCEDURE A BLOOD TYPING

1. Place a slide on a dry, flat surface.
2. Use a marking pencil to divide the slide into four sections and label the sections *H* (horse), *S* (sheep), *B* (bovine), and *U* (unknown) (Fig. 65-2).
3. Invert the blood and lectin bottle two or three times to be certain the materials are well mixed.
4. Have four different colored plastic microspatulas aligned and ready to be used for mixing.
5. Place one drop of lectin in the top of the first section of the slide. In the bottom of the first section place one drop of the designated blood. *Do not* allow the drops of the lectin and blood to touch each other at this time.

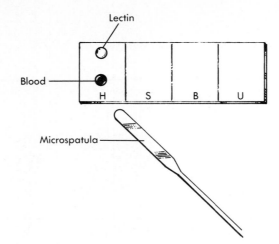

FIG. 65-2 Lectin and blood agglutination procedure.

Lectin

Blood

H S B U

Microspatula

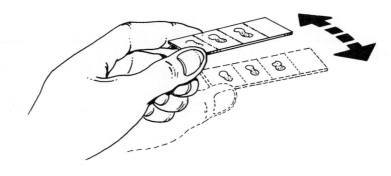

FIG. 65-3 Gently tilt slide back and forth.

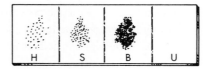

FIG. 65-4 Degrees of reaction.

H S B U

6. Use a separated microspatula to quickly mix the two drops in the first section together.
7. Gently tilt the slide back and forth (Fig. 65-3). Observe whether an agglutination reaction is taking place. Also observe the extent of the reaction. Classify the reaction (Fig. 65-4) as strong, moderate, weak, or none. **NOTE: Do not attempt to draw conclusions if more than 2 minutes have elapsed since the time you started mixing the blood and lectin. After this time, drying will begin and apparent results may not be accurate.**
8. Repeat steps 5 to 7 for each of the other bloods.
9. To identify the unknown, compare the reaction in the unknown circle with the others.

1. Explain what blood types A, B, AB, and O indicate.

2. Define the following terms:

   Agglutination

   Antigen

   Antibody

3. List the antibody or antibodies present in the plasma of A, B, AB, and O blood types.

4. Name the blood type of a universal recipient and a universal donor.

5. Explain the meaning of Rh+ and Rh− types.

6. Discuss the development of erythroblastosis fetalis.

*Use all references and materials at your disposal to answer these review questions.

66

LYMPHATIC SYSTEM
Gross anatomy and histology

OBJECTIVES

1 Identify the gross anatomy of the lymph nodes, tonsils, thymus, and spleen.
2 Identify the histological structure of the lymph nodes, tonsils, thymus, and spleen.
3 Explain the function of the lymph nodes, tonsils, thymus, and spleen.

MATERIALS

dissecting animal
dissecting equipment
dissecting microscope

histology slides
compound microscope

The **lymph vessels** (containing **lymph**), **lymph nodes** (glands), **tonsils, thymus,** and **spleen** are all part of the **lymphatic system.**

Lymphatic capillaries begin in the spaces between cells. These vessels are more permeable than blood capillaries. This allows the lymphatic system to drain any protein-containing fluid that cannot be reabsorbed after leaving the blood capillaries. The lymph vessels also function in the transportation of fats from the digestive tract to the blood, in the production of lymphocytes, and in the development of immunities.

Lymph nodes (Fig. 66-1 and Color Plate 17, *A*) are small oval structures of lymphoid tissue. Where blood vessels enter and lymphatic vessels leave is called the **hilum.** Lymph nodes are protected by a fibrous connective tissue called the **capsule.** A cross-section of a lymph node reveals an extension of the capsule called the **trabeculum,** an outer **cortex** with densely packed lymphocytes, and an inner **medulla** with lymphocytes arranged in strands. As lymph enters a node from the afferent vessel, it then circulates to the **medullary sinuses.** These vessels are lined with fixed agranular **phagocytic cells** that remove foreign substances. Most lymphocytes are formed in the lymph nodes.*

There are three types of tonsils: **pharyngeal,** in the posterior wall of the nasopharynx; **palatine,** in the tonsillar fossae; and **lingual,** at the base of the tongue. The tonsils also contain **phagocytic cells.**

The **spleen** (Fig. 66-2 and Color Plate 17, *B*) is protected by a **capsule** of fibroelastic tissue. Where blood vessels enter and lymph vessels leave is called the **hilum.** A cross-section reveals an extension of the capsule called the **trabeculum,** and outer **cortex,** and an inner **medulla.** Groups of lymphocytes surrounding arteries are called the **white pulp.** Blood-filled venous sinuses and stacks of splenic tissue are called the **red pulp.** The spleen

*Plasma cells are differentiated B cells that produce antibodies.

functions to phagocytize worn-out red blood cells and bacteria as well as to produce lymphocytes and plasma cells.

The **thymus** gland (Color Plate 17, *C*) processes lymphocytic stem cells into **T-cells** (thymus cells). The process allows cells to differentiate into cells that will provide specific immune reactions. Once this occurs, the T-cells migrate and become embedded in lymphoid tissue.

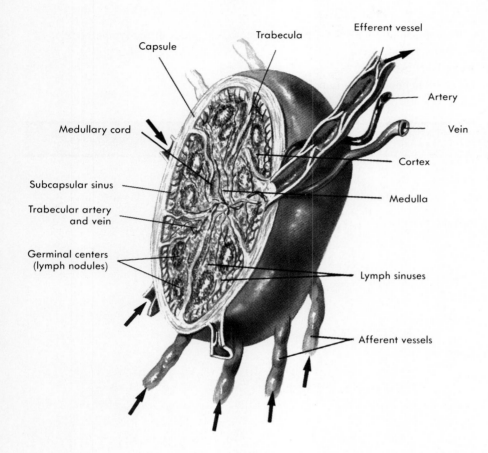

FIG. 66-1   Lymph node.

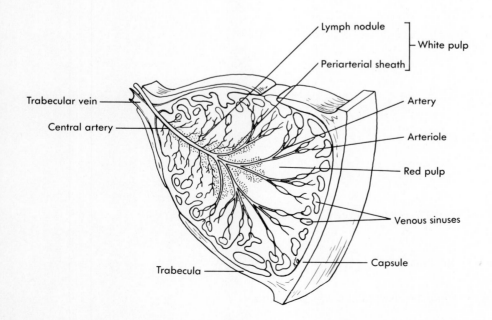

FIG. 66-2   Spleen section.

PROCEDURE A GROSS ANATOMY

1. Carefully remove any tissue or organ to locate a lymph node, the tonsils, the thymus, and the spleen.
2. Observe any external structures.
3. Make a cross-section of each and make observations with the dissecting microscope.

PROCEDURE B HISTOLOGY

1. Using the compound microscope, observe histological slides of the lymph node, tonsils, thymus, and spleen.
2. Draw and label each.
3. Explain the function of each lymphatic system structure.

REVIEW QUESTIONS*

1. Compare the structure of veins and lymphatic vessels.

2. Explain lymphatic circulation.

3. What is the function of the thymus gland?

4. Define the following terms:

   Lymphangiogram

   Edema

   Interferon

   Antigen

   Autoimmune

   Anaphylactic shock

   Antibody titer

5. Compare the various types of acquired immunity.

6. Draw, label, and explain the function of an antibody.

*Use all references and materials at your disposal to answer these review questions.

RESPIRATORY SYSTEM
Gross anatomy and histology

---

1 Identify the structures of the respiratory system.
2 Explain the function of the respiratory system.
3 Identify the histology of the various portions of the respiratory system.

KEY TERMS

Alveolar ducts
Alveoli
Arytenoid cartilage
Bronchi
Bronchioles
Corriculate cartilage
Cricoid cartilage
Cuneiform cartilage
Epiglottis
External respiration
Internal respiration
Larynx
Lungs
Nose
Parietal pleura
Pharynx
Pleural cavity
Pleural membrane
Primary bronchi
Respiratory bronchioles
Secondary bronchi
Septal cells
Squamous epithelial cells
Terminal bronchioles
Tertiary bronchi
Thyroid cartilage
Trachea
Ventilation
Visceral pleura

MATERIALS

dissecting animal
dissecting equipment
dissecting microscope

sterile drinking straw
respiratory system histology
slides

The cardiovascular and respiratory systems supply cells with oxygen and eliminate carbon dioxide. There are three processes in the respiration of humans: ventilation, external respiration, and internal respiration. During the process of **ventilation,** gases are exchanged between the atmosphere and lungs because of a pressure gradient. In **external respiration,** gases are exchanged between the lungs and blood. **Internal respiration** is the exchange of gases between the blood and body tissues.

The **respiratory organs** (Figs. 67-1 and 67-2) include the following:

Nose	Warms, moistens, and filters air
Pharynx	Passageway for air and food; provides resonance chamber for speech sounds; composed of skeletal muscles; lined with mucous membrane
Larynx	Voice box; contains three unpaired and three paired cartilages; cartilage components that may be observed easily include the following:
Epiglottis	Covers larynx during swallowing; prevents microbes from entering lower part of respiratory system
Thyroid cartilage	"Adam's apple"
Cricoid cartilage	Forms inferior wall of larynx
Arytenoid cartilage	Attaches vocal folds (true vocal cords) to pharyngeal muscles; action moves vocal cords
Corriculate cartilage	Accessory supporting laryngeal cartilages
Cuneiform cartilage	Accessory supporting laryngeal cartilages
Trachea	"Windpipe"; anterior to esophagus; horseshoe-shaped hyaline cartilage (Fig. 67-3) that provides support; open part of cartilage faces esophagus to provide expansion during swallowing; walls consist of smooth muscle and elastic connective tissue; epithelium consists of ciliated columnar cells, goblet cells, and basal cells

Bronchi	Horseshoe-shaped cartilage; epithelium consists of ciliated columnar cells; trachea, at distal end, bifurcates into **primary bronchi** (Fig. 67-4), **secondary bronchi, tertiary bronchi, bronchioles, terminal bronchioles,** and **respiratory bronchioles;** in the bronchioles, cartilage disappears and smooth muscle increases; terminal bronchiole epithelium contains simple cuboidal cells; respiratory bronchioles branch to form **alveolar ducts,** which lead to **alveoli** (Fig. 67-5); alveoli are made up of **squamous epithelial cells** and **septal cells;** gas exchange occurs across alveoli and capillary walls
Lungs	Right and left primary bronchi enter into right and left lung; right lung has three lobes and left lung has two lobes; **pleural membrane** (Fig. 67-6) consists of an outer layer called the **parietal pleura,** which attaches to the thoracic cavity, and an inner layer called the **visceral pleura,** which covers the lung tissue; the **pleural cavity,** formed by these membranes, contains fluid that prevents friction

FIG. 67-1 Upper respiratory system (sagittal section).

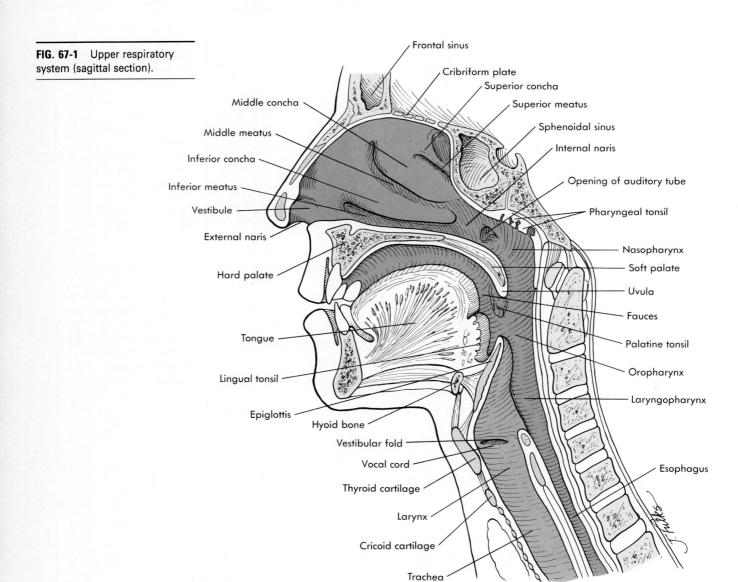

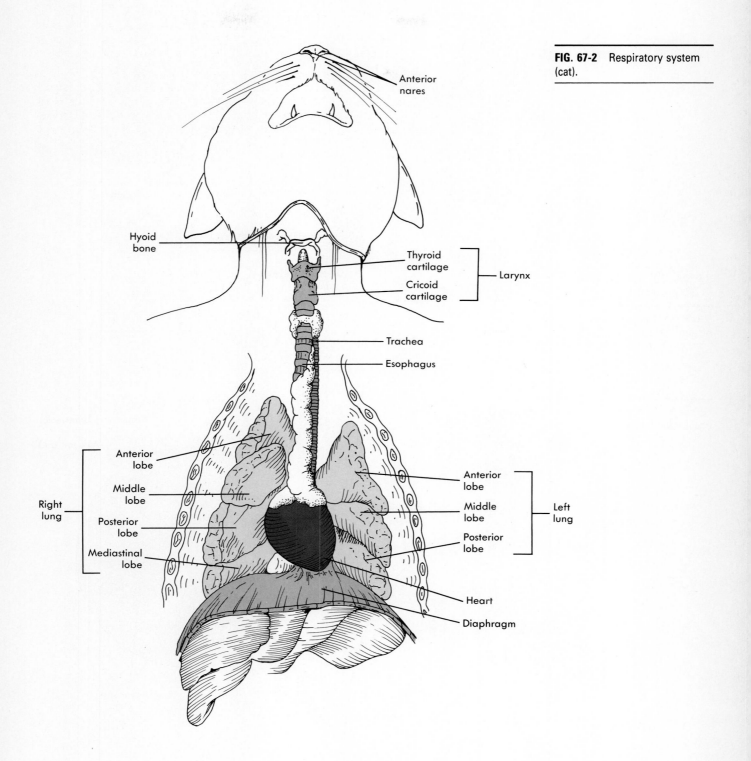

FIG. 67-2 Respiratory system (cat).

Anterior nares

Hyoid bone

Thyroid cartilage

Cricoid cartilage

Larynx

Trachea

Esophagus

Anterior lobe

Middle lobe

Posterior lobe

Mediastinal lobe

Right lung

Anterior lobe

Middle lobe

Posterior lobe

Left lung

Heart

Diaphragm

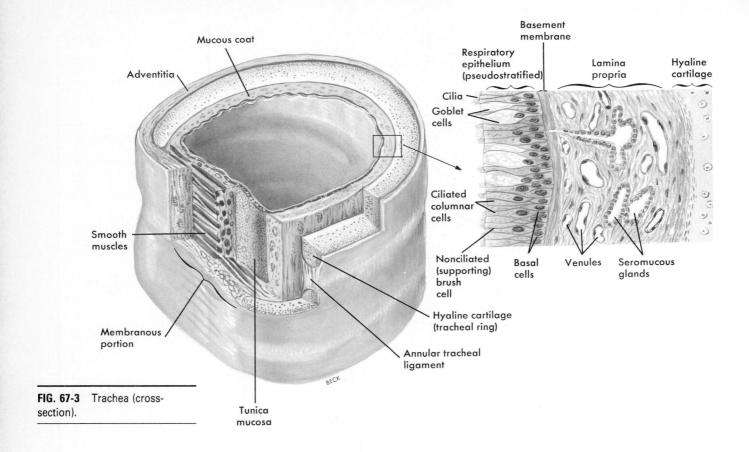

Adventitia

Mucous coat

Respiratory epithelium (pseudostratified)

Basement membrane

Lamina propria

Hyaline cartilage

Cilia

Goblet cells

Ciliated columnar cells

Nonciliated (supporting) brush cell

Basal cells

Venules

Seromucous glands

Smooth muscles

Membranous portion

Hyaline cartilage (tracheal ring)

Annular tracheal ligament

Tunica mucosa

BECK

FIG. 67-3 Trachea (cross-section).

FIG. 67-4 A, Anatomy of trachea and lungs.

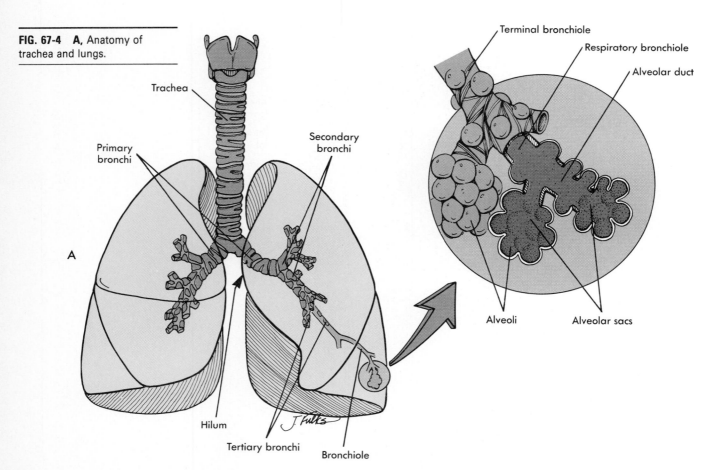

Trachea

Primary bronchi

Secondary bronchi

Terminal bronchiole

Respiratory bronchiole

Alveolar duct

A

Hilum

Tertiary bronchi

Bronchiole

Alveoli

Alveolar sacs

J Fulks

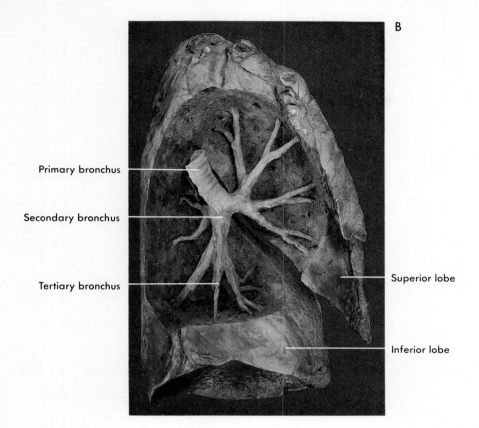

B

Primary bronchus

Secondary bronchus

Tertiary bronchus

Superior lobe

Inferior lobe

FIG. 67-4, cont'd   B, Bronchiole distribution.

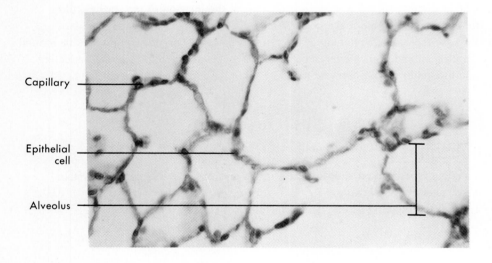

Capillary

Epithelial cell

Alveolus

FIG. 67-5 Alveoli histology.

FIG. 67-6   Pleural membrane.

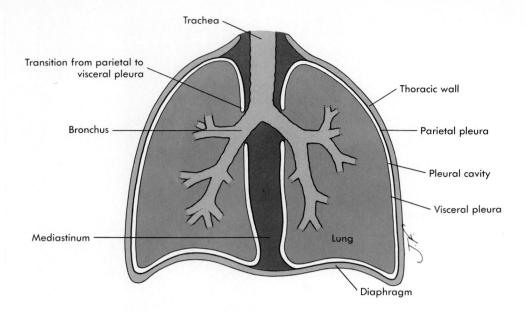

PROCEDURE A   EXAMINATION OF THE RESPIRATORY SYSTEM

1. To remove the larynx, make a cut just above the epiglottis and along the trachea.
2. Lift the epiglottis and note the vocal folds.
3. Use the dissecting microscope to identify the cartilage components of the larynx.
4. Remove a small section of the trachea; use the dissecting microscope to observe the cartilage.
5. Obtain a sterile drinking straw and insert it in the incision of the trachea; blow through the straw and make observations.
6. Remove the lungs; beginning at the primary bronchi, *carefully* remove tissue around the bronchi and bronchioles (see Fig. 67-4) as far as possible. Use the dissecting microscope to make observations.

PROCEDURE B   HISTOLOGY

1. Examine cross-section slides of the various portions of the respiratory system.
2. Draw and label each slide.
3. Explain the function of the various histological structures of each slide.

REVIEW QUESTIONS*

1. Distinguish between the upper and lower respiratory system.

2. Why is the location of the epiglottis important?

3. Explain how respiratory gases are transported by the blood.

4. What is the function of the paranasal sinuses?

5. What is the cause of pleurisy?

6. Why is the right lung larger than the left?

7. Explain how voice sound is produced.

8. What causes laryngitis?

*Use all references and materials at your disposal to answer these review questions.

68

RESPIRATORY SYSTEM
Lung capacity

KEY TERMS

Expiration
Expiratory reserve volume
Inspiration
Inspiratory reserve volume
Residual volume
Respiration
Respirometer
Spirometer
Tidal volume
Vital capacity

OBJECTIVES

1 Demonstrate and explain lung capacity.

MATERIALS

balloon   ruler

FIG. 68-1 Lung volumes and capacities.

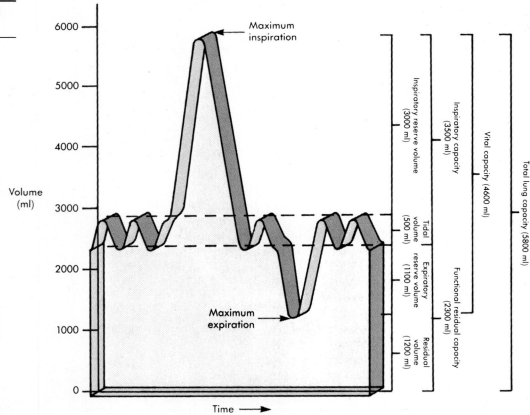

Inspiration, breathing in, and expiration, breathing out, occur because of a pressure gradient. Clinically, the term respiration includes one inspiration plus one expiration.

The amount of air taken in or expelled during normal breathing is about 500 ml. This volume of air is called the tidal volume (Fig. 68-1). When a very deep breath is taken, the volume of inhaled air averages about 3,100 ml above normal tidal volume. This is called the inspiratory reserve volume. When a normal breath is taken then forcibly expelled, the volume of exhaled air averages about 1,200 ml above normal tidal volume. This is called the expiratory reserve volume. The largest possible amount of air that can be expelled after taking in a deep breath, inspiratory reserve volume plus expiratory reserve volume, is the vital capacity. Lower intrathoracic pressure allows some air to remain in the lungs. This volume is about 1,200 ml and is called the residual volume.

Human lung capacity can be measured in several ways. One way is by using a piece of laboratory equipment called a respirometer or spirometer. A less accurate way to measure lung capacity is by using a balloon.

PROCEDURE A VITAL CAPACITY

1. Stretch a balloon several times.
2. Take as deep a breath as possible; exhale as much air as possible into the balloon; pinch the balloon closed to prevent air from escaping.
3. Measure and record the diameter of the balloon in centimeters.
4. Repeat this procedure four more times; record each diameter.
5. Record the vital capacity average.

PROCEDURE B EXPIRATORY RESERVE

1. Inhale and exhale normally.
2. Now exhale all the air still left into the balloon.
3. Measure and record the diameter of the balloon in centimeters.
4. Repeat this procedure four more times; record each diameter.
5. Record the expiratory reserve volume average.

PROCEDURE C TIDAL VOLUME

1. Take in a normal breath; exhale into the balloon only as much air as you would normally exhale. *Do not force* the air.
2. Measure and record the diameter of the balloon in centimeters.
3. Repeat this procedure four more times; record each diameter.
4. Record the tidal volume average.

PROCEDURE D CONVERSION OF DIAMETER TO VOLUME

1. Lung volume is expressed in cubic centimeters (cm^3).
$$1,000 \ cm^3 = 1,000 \ ml \ (1 \ L)$$
2. To convert from balloon diameter to volume, locate the balloon diameter on the horizontal axis of the provided graph. Follow this number up to the dark, then move across to locate the corresponding volume (Fig. 68-2).
3. Convert each diameter for vital capacity, expiratory reserve, and tidal volume to volume.

FIG. 68-2 Lung volume.

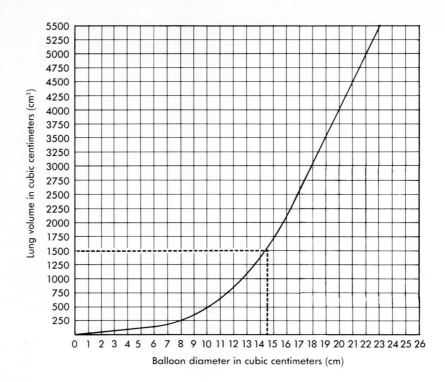

Lung volume in cubic centimeters (cm³)

Balloon diameter in cubic centimeters (cm)

REVIEW QUESTIONS*

1. Compare your averages with the following averages taken by a spirometer.

	MALE	FEMALE
Vital capacity	5,000 ml	4,000 ml
Expiratory reserve	1,200 ml	1,000 ml
Tidal volume	525 ml	475 ml

2. What is your normal breathing rate per minute?

3. How much air do you inhale in 1 minute?

4. Explain the following disorders:

   Bronchial asthma

   Emphysema

   Pneumonia

   Tuberculosis

*Use all references and materials at your disposal to answer these review questions.

DIGESTIVE SYSTEM
Gross anatomy

69

OBJECTIVES

1. Explain the function of the mouth and its accessory organs.
2. Describe the mechanical digestion associated with the pharynx and esophagus.
3. Discuss the structural features of the stomach and how it relates to mechanical and chemical digestion.
4. Explain how the structure of the small intestine relates to the function.
5. Discuss the relationship of the liver, pancreas, and gallbladder to the small intestine.
6. Describe the function of the large intestine.
7. Explain the function of the mesentery, falciform ligament, greater omentum, and lesser omentum.

KEY TERMS

Acinar gland
Acini
Alimentary canal
Ampulla of Vater
Amylase
Anal canal
Bile
Body
Bolus
Brunner's glands
Cardia portion
Cecum
Chief cells
Chyme
Colon
Common bile duct
Common hepatic duct
Crypts of Lieberkühn
Cystic duct
Deglutition
Digestive system
Duct of Wirsung
Duodenum
Esophagus
Falciform ligament
Fundus
Gallbladder
Gastric glands
Gastroesophageal sphincter
Gastrointestinal tract
Glucagon
Greater curvature
Greater omentum
Hepatic cells
Hydrochloric acid
Ileocecal valve
Ileum
Insulin
Islets of Langerhans
Jejunum
Large intestine

MATERIALS

dissecting animal   dissecting microscope
dissecting equipment

The **digestive system** (Figs. 69-1 and 69-2) may be divided into two groups: the **gastrointestinal (GI) tract,** or **alimentary canal,** and the **accessory organs.** The alimentary canal includes the **mouth, pharynx, esophagus, stomach, small intestine,** and **large intestine.** The muscles in the walls along these organs physically break down ingested food, and secretions produced by the cells of these organs chemically break down the food. The accessory organs include the **teeth, tongue, salivary glands, gastric and intestinal glands, liver, gallbladder, pancreas,** and **appendix.** The tongue and teeth aid in the physical breakdown of food. The remaining accessory organs produce or store secretions released through ducts into the GI tract for the chemical breakdown of food.

The physical structure and chemical or physical function of the GI tract and accessory organs are as follows:

Mouth	During chewing, food is kept between upper and lower teeth with cheeks and lips: **tongue** voluntary muscles maneuver food for chewing **(mastication)** and swallowing **(deglutition)**
Teeth	Used for pulverizing food
Salivary glands (Fig. 69-3)	Secrete saliva: **water** components dissolve foods for tasting and chemical reactions; **mucin** aids in lubrication of food; **amylase,** digestive enzyme, provides chemical digestion; **lysozyme** aids in destruction of bacteria
Parotid glands	Located between skin and masseter muscles; **tubuloacinar glands** secrete saliva into mouth through **Stensen's duct**

Continued.

FIG. 69-1 Digestive system.

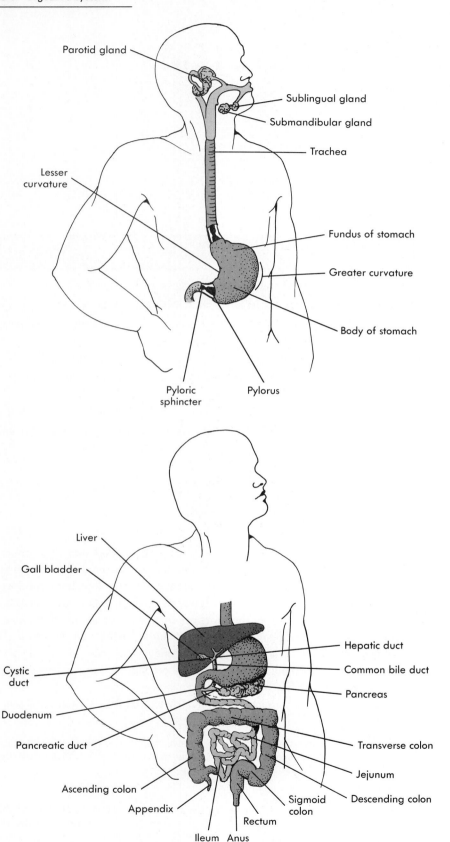

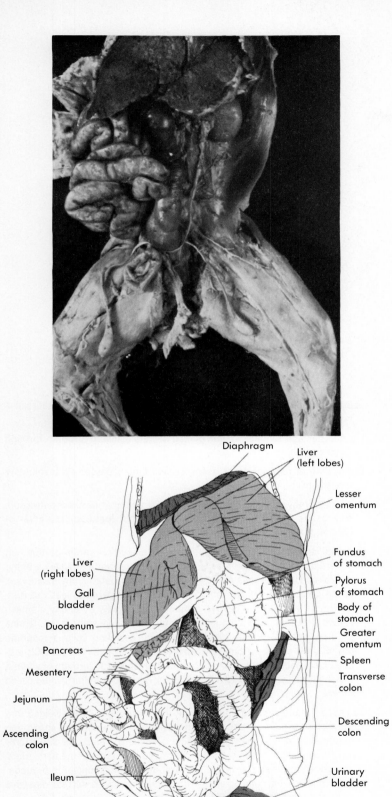

FIG. 69-2 Digestive system (cat).

Diaphragm

Liver
(left lobes)

Lesser
omentum

Liver
(right lobes)

Fundus
of stomach

Gall
bladder

Pylorus
of stomach

Duodenum

Body of
stomach

Pancreas

Greater
omentum

Mesentery

Spleen

Jejunum

Transverse
colon

Ascending
colon

Descending
colon

Ileum

Urinary
bladder

FIG. 69-3 Salivary glands.

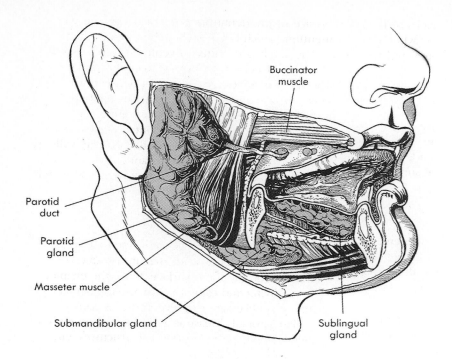

Submandibular glands	Located beneath and posterior to base of tongue; **acinar glands** secrete saliva into mouth through **Wharton's duct**
Sublingual glands	Located beneath tongue; **acinar glands** secrete saliva into mouth through direct ducts
Pharynx (throat)	Swallowing of rounded mass of food called **bolus;** passes via involuntary movement
Esophagus	Involuntary movement of bolus by **peristaltic action,** muscular contraction, and glands secreting mucus; **gastroesophageal (cardiac) sphincter** at inferior part leading into stomach
Stomach	J-shaped structure; below gastroesophageal sphincter is **cardia** portion; left of cardia is curved portion called **fundus;** central portion is called **body;** medial concave curve is called **lesser curvature;** lateral convex curve is called **greater curvature;** inferior portion, called **pylorus,** leads to duodenum of intestine via **pyloric sphincter;** internally, an empty stomach shows large folds called **rugae;** in full stomach, rugae smooth out; lamina propria contains **gastric glands** lined with secreting cells: **zymogenic (chief) cells** secrete enzyme **pepsinogen; parietal cells** secrete **HCl; mucous cells** secrete mucus and substance that aids in adsorption of vitamin B_{12}; ingested food is mixed with combined secretions to form liquid called **chyme**
Small intestine	So called because of small diameter of this long folded tube; divided into three sections: **duodenum, jejunum,** and **ileum;** its function is absorption of nutrients; **villi,** fingerlike projections of mucous membrane, and **microvilli,** projections of each cell's plasma membrane, provide greater surface area for absorption; chyme passes through pyloric sphincter into duodenum of small intestine; glands within mucosa, which secrete digestive enzymes, are called **crypts of Lieberkühn; Brunner's glands** lie in submucosa and secrete alkaline mucus to neutralize acidic chyme as well as protect walls from digestive enzymes

    The small intestine depends on chemical digestion from its own cells and cells located in the following accessory organs:

Pancreas	Located posterior to greater curvature of stomach; connected to duodenum of small intestine by duct; contains clusters of endocrine cell called **islets of Langerhans,** which secrete **insulin** and **glucagon** to regulate blood sugar level; clusters of exocrine cells called **acini** secrete digestive enzymes,

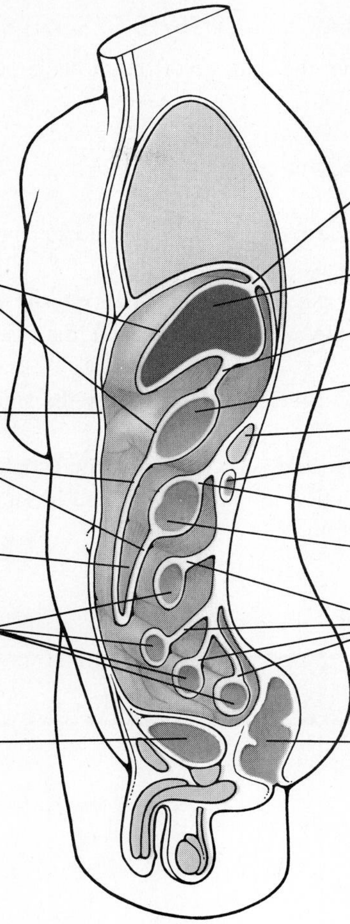

FIG. 69-4 Peritoneum and mesenteries associated with some abdominal organs (sagittal section).

Labels in figure:
Coronary ligament
Visceral peritoneum
Liver
Lesser omentum
Stomach
Parietal peritoneum
Pancreas
Duodenum
} Retroperitoneal
Greater omentum
Transverse mesocolon
Transverse colon
Omental bursa
Small intestine
Mesentery proper
Urinary bladder (retroperitoneal)
Rectum (retroperitoneal)

which pass into the **duct of Wirsung; joins common bile duct** to form **ampulla of Vater;** this duct empties into duodenum of small intestine

Liver   Located under diaphragm and made up of two lobes; **hepatic cells** secrete **bile,** which is used for emulsification and absorption of fats and passed through a **common hepatic duct;** joins **cystic duct** from gallbladder to form common bile duct

Gallbladder   Small sac located under liver; inner wall has rugae, which allow it to fill with bile; when small intestine is empty, **sphincter of Oddi** (of the common bile duct) closes, and excess bile overflows through cystic duct into gallbladder; bile is stored here until needed

The remaining portion of the alimentary canal is the following:

Large intestine   Final portion of small intestine, **ileum,** leads into large intestine by way of **ileocecal valve;** large intestine is divided into **cecum, vermiform appendix, colon, rectum,** and **anal canal;** absorption of nutrients is completed, excess water is absorbed, vitamins are synthesized, and feces is formed and eliminated

Important to the digestive system is the **peritoneum** (Fig. 69-4), a **serous membrane** with simple squamal epithelial tissue and connective tissue. **Visceral peritoneum** covers organs and is called the **serosa.** An extension of the peritoneum, called the **mensentery,** binds the small intestine to the abdominal wall. Another extension of the peritoneum, the **mesocolon,** binds the large intestine to the posterior wall. The **falciform ligament** attaches the liver to the diaphragm and abdominal wall. Also extensions of the peritoneum are the **lesser omentum** and **greater omentum.** The lesser omentum attaches the liver to the stomach and the duodenum. The greater omentum is a peritoneal extension of the stomach that lies over the intestines and attaches to the posterior wall of the abdominal cavity.

PROCEDURE A   GROSS ANATOMY OF THE DIGESTIVE SYSTEM

1. Make a longitudinal incision from the midline of the thoracic cavity up to the oral cavity.
2. Remove any tissue to locate and identify the salivary glands.
3. Follow the alimentary canal along the pharynx and esophagus, through the diaphragm, to the stomach.
4. Lying on top of the stomach is the liver; note the gallbladder between the liver lobes.
5. Lift the liver and note the pancreas.
6. Follow the ducts of the liver and pancreas to the small intestine.
7. Remove the omentum to locate the small intestine; note the mesentery and its vascularity.
8. Note where the stomach and small intestine join; follow the small intestine to the large intestine; note the mesocolon.
9. Remove a section along each digestive structure from the esophagus to the large intestine; make a longitudinal section in each section and observe how the structural characteristics relate to the function.

REVIEW QUESTIONS*

1. Explain the muscular movements of the small intestine.

2. Explain the function of the following pancreatic enzymes:

   Trypsin
   Chymotrypsin
   Carboxypeptidase
   Pancreatic amylase
   Pancreatic lipase
   Nuclease

3. Explain the function of the salivary glands in digestion.

4. Define the following terms:

   Mastication
   Bolus
   Deglutition
   Peristalsis
   Sphincter

*Use all references and materials at your disposal to answer these review questions.

5. Discuss the function of the following structures:

   Gastroesophageal sphincter
   Pyloric valve
   Ileocecal valve

6. Explain the function of the pancreas, liver, and gallbladder.

7. Describe the following disorders:

   Peptic ulcer
   Gallstones
   Appendicitis
   Diverticulitis
   Peritonitis

8. Describe the extensions of the peritoneum in association with the alimentary canal.

70 DIGESTIVE SYSTEM
Histology

KEY TERMS

Alimentary canal
Epithelial layer
Lamina propria
Muscularis mucosa
Plexus of Auerbach
Plexus of Meissner
Tunica mucosa
Tunica muscularis
Tunica serosa
Tunica submucosa
Visceral peritoneum

OBJECTIVES

1 Identify the wall of the alimentary canal.
2 Explain the function of the tunics of the alimentary canal.

MATERIALS

compound microscope   alimentary canal prepared slides

Histologically, the **alimentary canal** has four layers or tunics: **mucosa** (inner layer), **submucosa, muscularis,** and **adventitia** (serosa) (Fig. 70-1). Each layer may be characterized as follows:

Tunica mucosa	Inner lining with sublayers:
Epithelial layer	Contains nonkeratinized cells; stratified in mouth, esophagus, and anus for protection and secretion; simple in rest of tract for secretion and absorption
Lamina propria	Highly vascular loose connective tissue; lymph tissue provides protection from disease; contains glands that secrete products for chemical digestion
Muscularis mucosa	Contains visceral muscle for involuntary movement
Tunica submucosa	Areolar connective tissue attaching mucosa and muscularis layers; highly vascular; contains autonomic nerve supply called the **plexus of Meissner**
Tunica muscularis	Skeletal muscle in mouth, pharynx, and esophagus; smooth muscle throughout rest of tract; arranged as inner circular layer and outer longitudinal layer; third layer is found in stomach; provides peristaltic action for physical digestion; contains autonomic nerve supply called **plexus of Auerbach**
Tunica serosa	Outer layer of connective tissue and epithelium called serous membrane; also called **visceral peritoneum**

PROCEDURE A HISTOLOGY

1. Observe the tunica mucosa, tunica submucosa, tunica muscularis, and tunica serosa of the alimentary canal under high power. Compare the function of each.
2. Draw and label a cross-section of the alimentary canal.

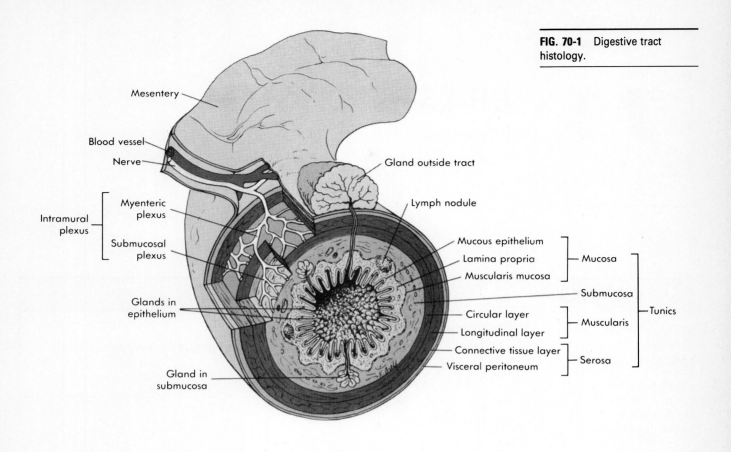

FIG. 70-1 Digestive tract histology.

Mesentery

Blood vessel

Nerve

Gland outside tract

Lymph nodule

Intramural plexus
- Myenteric plexus
- Submucosal plexus

Glands in epithelium

Mucous epithelium
Lamina propria
Muscularis mucosa
}— Mucosa

Submucosa

Circular layer
Longitudinal layer
}— Muscularis

Connective tissue layer
Visceral peritoneum
}— Serosa

}—Tunics

Gland in submucosa

REVIEW QUESTIONS*

1. Explain the function of the tunica mucosa sublayers:

   Epithelial layer
   Lamina propria
   Muscularis mucosa

2. Discuss the purpose of the tunica submucosa.

3. How is the tunica muscularis important to the alimentary canal?

4. Draw and label the four tunics of the alimentary canal.

5. Explain the function of the plexus of Meissner and plexus of Auerbach.

*Use all references and materials at your disposal to answer these review questions.

DIGESTIVE SYSTEM
Physiology

KEY TERMS

Bile salt
Chief cells
Hydrochloric acid
Lipase
Litmus
Pepsin
Pepsinogen
Salivary amylase

OBJECTIVES

1   Discuss the digestion of carbohydrates.
2   Explain the relationship between bile salts and pancreatic enzymes in fat digestion.
3   Explain protein digestion in the stomach.

MATERIALS

3 test tubes	test tube rack
distilled water	hot plate
cream	beaker
droppers	sodium hydroxide
litmus	amylase
hydrochloric acid	1% boiled Benedict's reagent
casein	beaker
lipase	pepsinogen
bile	hydrion paper

Chemical digestion of fats, proteins, and carbohydrates occurs because of the action of enzymes produced by organs of the digestive system.

The parotid, submandibular, and sublingual glands produce **salivary amylase (ptyalin),** an enzyme that hydrolyzes cooked starch and glycogen to maltose. In the presence of a reducing sugar, such as maltose, Benedict's solution, an alkaline copper sulfate reagent with sodium citrate, will be reduced to a red cuprous oxide. The amount of reducing sugar in a solution will determine the color reaction (ranging from green, yellow, orange, or red) with the Benedict's reagent.

Pancreatic enzymes, called **lipases,** can catabolize fats (triglycerides) into fatty acids and glycerols. The effect of lipase is greatly increased when the fats are emulsified by **bile salts. Litmus,** a pH color indicator, is blue in the presence of fat. As fat is digested to fatty acids and glycerol, the acidic environment may be detected as the litmus changes to pink.

Specialized cells in the lining of the stomach, called **chief cells,** secrete the proteolytic enzyme, **pepsin.** Pepsin is secreted in the inactive form, **pepsinogen,** and is activated by **hydrochloric acid (HCl)** secreted by parietal cells lining the stomach.

Once chemical digestion of food occurs, the nutrients are absorbed into the blood system, primarily through the walls of the small intestine.

PROCEDURE A CARBOHYDRATE DIGESTION

1. Label two test tubes and add the following solutions:

   Tube 1 3 ml boiled starch solution

           3 ml Benedict's reagent

           3 ml distilled water

   Tube 2 3 ml boiled starch solution

           3 ml Benedict's reagent

           3 ml amylase

2. Place test tubes in a boiling water bath for 2 minutes.
3. Record results.

PROCEDURE B FAT DIGESTION

1. Label three test tubes and add the following solutions:

   Tube 1 10 ml distilled water

           10 ml litmus

            1 drop cream

   Tube 2 10 ml distilled water

           10 ml litmus

            1 drop cream

            5 ml lipase

   Tube 3 10 ml distilled water

           10 ml litmus

            1 drop cream

            5 ml lipase

            3 ml bile

2. Mix the contents by swirling each test tube.
3. Place the test tubes in a rack; swirl the test tubes every 10 minutes. After 30 minutes record any color change.

PROCEDURE C PROTEIN DIGESTION

1. Label 5 test tubes and add the following solutions:

   Tube 1 5 ml 0.5% pepsin

           5 ml distilled water

   Tube 2 5 ml 0.5% pepsin

           5 ml 0.8% HCl

   Tube 3 5 ml 0.5% pepsin (boiled for 10 minutes in a water bath)

           5 ml 0.8% HCl

   Tube 4 5 ml 0.5% pepsin

           5 ml 0.5% NaOH

   Tube 5 5 ml distilled water

           5 ml 0.8% HCl

2. Use hydrion paper to determine the approximate pH of each test tube.
3. Place a piece of casein in each test tube and then place the test tubes in a 40° C water bath, swirling the test tubes from time to time.
4. Continue until one of the casein pieces becomes transparent and gradually dissolves.
5. Record observations.

1. What is the function of the salivary amylase?

2. Discuss the different substances that each salivary gland supplies to saliva.

3. Explain how the following specialized cells of the stomach contribute to chemical digestion.

   Chief (zymogenic) cells
   Parietal cells
   Mucous cells

4. Why are protein digestive enzymes secreted in an inactive form?

5. Explain the importance of the digestive enzymes renin and lipase.

6. What is the importance of bile in fat digestion?

7. Explain how litmus may be used to detect fat digestion.

*Use all references and materials at your disposal to answer these review questions.

URINARY SYSTEM
Gross anatomy and histology

72

OBJECTIVES

1 Identify the structures of the urinary system.
2 Explain the function of each structure of the urinary system.

KEY TERMS

Adipose capsule
Afferent arteriole
Ascending limb of Henle
Bowman's capsule
Collecting duct
Cortical nephron
Descending limb of Henle
Distal convoluted tubule
Efferent arteriole
Glomerulus
Hilum
Juxtamedullary nephron
Kidney
Loop of Henle
Major calyx
Minor calyx
Nephron
Proximal convoluted tubule
Renal artery
Renal capsule
Renal fascia
Renal pelvis
Renal vein
Renal venule
Urethra
Ureter
Urinary bladder
Urinary system
Vasa recta

MATERIALS

dissecting animal
dissecting equipment
dissecting microscope

prepared slides
compound microscope

The **urinary system,** along with the integumentary, respiratory, and digestive systems, eliminates bodily waste. The urinary system also controls homeostasis of the body by regulating composition and volume of blood as well as removing cellular metabolite by-products from the blood in the form of urine.

The urinary system consists of two **kidneys,** two **ureters,** a **urinary bladder,** and a **urethra** (Fig. 72-1, *A* and *B*). The kidneys are located posterior to the peritoneum *(retroperitoneal)* of the abdominal cavity. They are surrounded by three layers of tissue. The outermost layer is the **renal fascia,** which consists of a thin layer of connective tissue that attaches the kidney to the abdominal wall. The **adipose capsule** is a fatty tissue layer that functions in protection. The inner layer is the **renal capsule** (Fig. 72-2, A), which consists of a transparent fibrous membrane that isolates the kidney.

Each kidney consists of approximately 1 million tubular structures called **nephrons.** They may be classified as either **cortical nephrons** or **juxtamedullary nephrons.**

Nephrons (see Fig. 72-2, B) begin with a double-walled cuplike structure called **Bowman's capsule.** The cuplike structure contains a capillary network called the **glomerulus.** Continuing from the Bowman's capsule is a series of tubules. The **proximal convoluted tubule** (PCT) suggests its location is nearest to the Bowman's capsule. The PCT continues to form the **loop of Henle,** which consists of the **descending limb of Henle.** A structure called the **hairpin loop of Henle** continues to form the **ascending limb of Henle.** The ascending limb of Henle then gives rise to the **distal convoluted tubule** (DCT). The DCT leads to a **collecting duct, minor calyx, major calyx, renal pelvis,** and then the ureter.

The **hilum** contains the ureter, blood vessels, and lymphatics. Blood enters the kidneys by way of the **renal artery,** which branches into the lobular arteries, arcuate arteries, and interlobular **afferent arterioles.** The afferent arteriole then gives rise to about 50 capillary loops collectively called the glomerulus. The glomerular capillaries then join and form the **efferent arteriole.** The efferent arteriole thins to produce **peritubular capillaries** that surround the PCT, loop of Henle, and DCT in both types of nephrons. The peritubular capillaries of the juxtaglomerular nephrons are called the **vasa recta.** The vessels lead to **renal venules** and to the **renal vein,** which exits at the hilum.

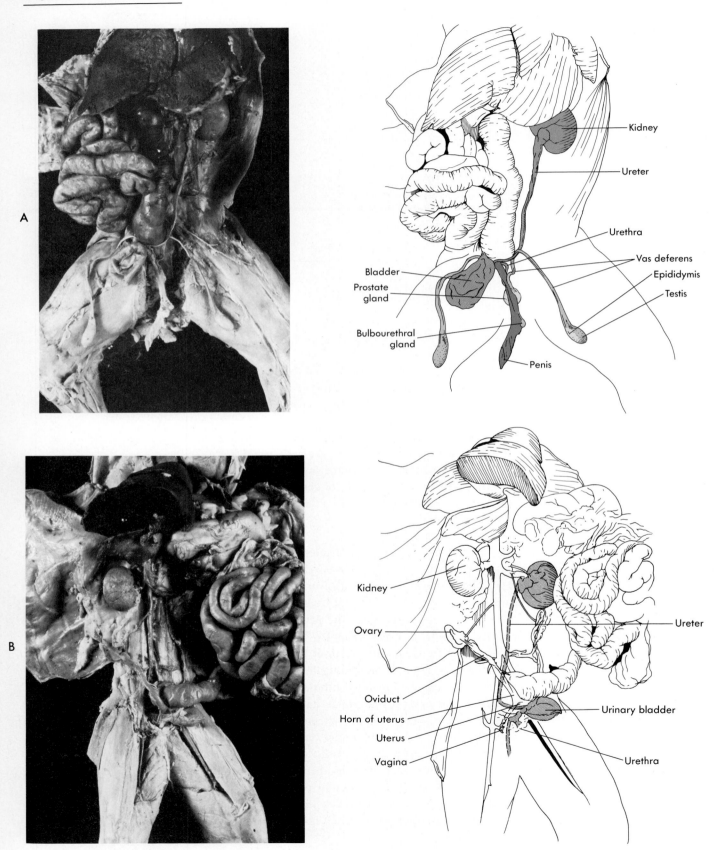

Kidney

Ureter

Urethra

Bladder

Vas deferens

Prostate gland

Epididymis

Testis

Bulbourethral gland

Penis

Kidney

Ovary

Ureter

Oviduct

Horn of uterus

Urinary bladder

Uterus

Vagina

Urethra

A

B

FIG. 72-2 A, Longitudinal
section of kidney and ureter.

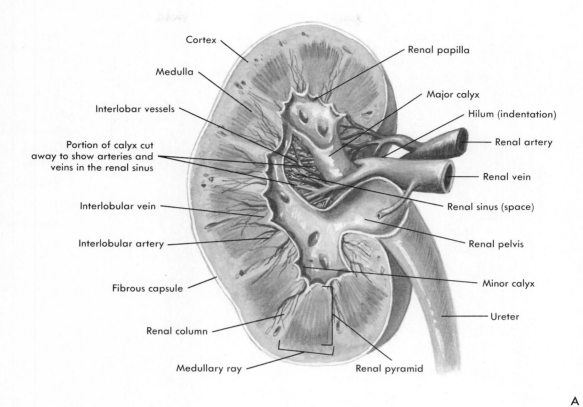

Cortex

Renal papilla

Medulla

Major calyx

Interlobar vessels

Hilum (indentation)

Renal artery

Portion of calyx cut
away to show arteries and
veins in the renal sinus

Renal vein

Interlobular vein

Renal sinus (space)

Interlobular artery

Renal pelvis

Fibrous capsule

Minor calyx

Ureter

Renal column

Medullary ray

Renal pyramid

A

Renal column

Renal pyramids

Adrenal gland

Minor calices

Renal pelvis

Renal vein

Renal artery

Major calyx

Ureter

Fibrous capsule

Continued.

FIG. 72-2, cont'd B,
Longitudinal section of kidney
and ureter.

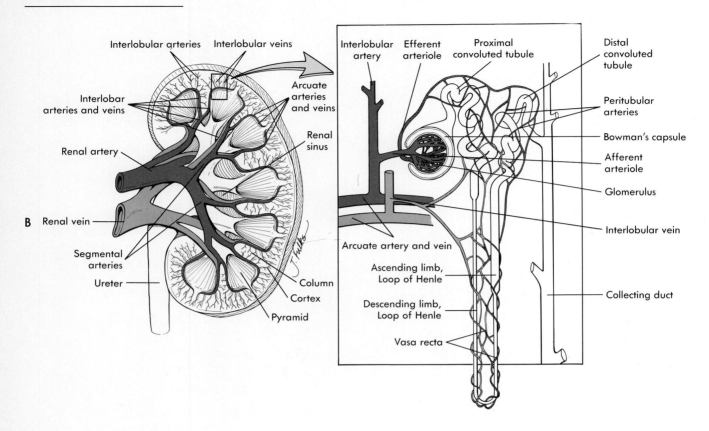

B

<u>**PROCEDURE A**</u>   GROSS ANATOMY OF THE KIDNEY

1. Carefully lift any organs to locate the kidneys behind the abdominal cavity; remove
   any peritoneum; note the renal fascia, adipose capsule, and renal capsule.
2. Locate the hilum and identify the ureter, renal artery, and renal vein.
3. Follow the ureter to the urinary bladder and then to the urethra.
4. Remove a kidney and cut to separate the anterior and posterior portions.
5. Using the dissecting microscope, identify the internal structures.

<u>**PROCEDURE B**</u>   HISTOLOGY OF THE KIDNEY

1. Identify the structures of a nephron using histological slides.
2. Draw and label your observations.
3. Explain the function of each structure.

1. Draw and label the internal gross anatomy of the kidneys.

2. Describe the blood and nerve supply to the kidneys.

3. List the normal chemical components of urine.

4. Define the following terms:
   Glomerulonephritis
   Ketosis
   Bilirubinuria
   Casts
   Renal calculi

5. Explain the micturition reflex.

6. Describe the process of urine formation.

7. Describe the structure and function of the juxtaglomerular apparatus.

8. Explain tubular secretion and tubular reabsorption.

*Use all references and materials at your disposal to answer these review questions.

73 URINARY SYSTEM
Physiology

KEY TERMS

Acetic acid
Albumin
Benedict's reagent
Chlorides
Glucose
Phosphates
Silver nitrate
Urea
Urine

OBJECTIVE

1 Demonstrate the clinical procedures in detecting the presence of glucose, chloride, albumin, and phosphate.

MATERIALS

test tubes
Bunsen burner
test tube holder
beaker
hot plate
Benedict's reagent

glucose solution
phosphate solution
albumin solution
silver nitrate*
acetic acid

* CAUTION: if spillage occurs
rinse well with water

The kidneys function in maintaining a normal balance of specific substances within the body. **Urine** is made of wastes from the body. These wastes include a poisonous substance called **urea** as well as excess salts, sugar, and other chemicals. Clinical tests indicate the chemical composition of urine. The presence or absence of certain chemicals may indicate whether or not certain body processes are occurring.*

Some of the clinical tests performed include testing for excess glucose, chloride, albumin, and phosphates in the urine. **Glucose** and **Benedict's reagent,** in the presence of heat, will change from blue to green, orange, or red to indicate varying amounts of sugar. If **chlorides** are present in urine, a white cloud (precipitate) will form when **silver nitrate** is added. To test for the presence of **albumin,** a test tube is heated at the top of the urine. A haze will then form. If **acetic acid** is added and the haze remains, albumin is present. The test for **phosphates** in urine is similar to albumin. If a test tube is heated along the top of the urine, a haze will form; however, if acetic acid is added and the haze disappears, phosphate is present.

*For health reasons, all urine samples in these procedures are synthetic.

PROCEDURE A GLUCOSE IDENTIFICATION

1. Prepare a hot water bath.
2. To a properly labeled test tube, add 3 ml of known glucose/urine sample and 3 ml of Benedict's reagent.
3. Place the test tube in a hot water bath for 5 minutes; record the results.

PROCEDURE B CHLORIDE IDENTIFICATION

1. To a properly labeled test tube, add 5 ml of known chloride/urine sample.
2. Add several drops of silver nitrate.
3. Record the results.

PROCEDURE C ALBUMIN IDENTIFICATION

1. To a properly labeled test tube, add 10 ml of known albumin/urine sample.
2. Using a Bunsen burner, heat only the top portion of the urine sample.
3. Add 4 to 5 drops of acetic acid.
4. Record the results.

PROCEDURE D PHOSPHATE IDENTIFICATION

1. To a properly labeled test tube, add 10 ml of known phosphate/urine sample.
2. Using a Bunsen burner, heat only the top portion of the urine sample.
3. Add 4 to 5 drops of acetic acid.
4. Record the results.

PROCEDURE E UNKNOWN IDENTIFICATION

1. Obtain a flask of an unknown urine sample.
2. Add the appropriate amount of sample to four properly labeled test tubes: one each for glucose, chloride, albumin, and phosphate.
3. Perform the appropriate tests and record the results.

1. Define the following terms:

   Pyelitis
   Pyelonephritis
   Cystitis
   Polycystic disease
   Albuminuria

2. Explain the operational principle of hemodialysis.

3. Compare the effects of blood pressure, blood concentration, temperature, diuretics, and emotions on urine volume.

4. Are fats found in urine? Why or why not?

*Use all references and materials at your disposal to answer these review questions.

REPRODUCTIVE SYSTEM
Male reproductive system

OBJECTIVES

1 Name the structures of the male reproductive system.

2 Explain the function of the structures of the male reproductive system.

3 Discuss the movement of a sperm cell from the testes through a series of ducts to the exterior.

MATERIALS

dissecting animal (male)   testes prepared slide
dissecting equipment   compound microscope

The **gonads,** or **testes,** of the **male reproductive system** function to produce **gametes,** or **sperm cells,** and secrete hormones. **Ducts** of the reproductive system transport, receive, and store gametes. **Accessory glands** in the reproductive system secrete materials that support gametes. The male reproductive system structures (Fig. 74-1 and Color Plate 18) function as follows:

Scrotum	Outpouching of loose skin and superficial fascia of abdominal wall; supporting structure for testes
Testes	Covered by dense connective tissue called **tunica albuginea,** which divides testis into separate **lobules** or sections; each lobule contains coiled **seminiferous tubules** (Fig. 74-2) lined by **germinal epithelium,** which produces **spermatozoa** by process of **spermatogenesis;** along basement membrane are immature cells that mature as they move toward lumen; immature cells are **spermatogonia** and mature into **primary spermatocytes, secondary spermatocytes,** and then **spermatids;** full maturity of cell is reached in lumin of reproductive ducts; also found in germinal epithelium of **Sertoli cells,** which secrete nutrients for spermatozoa; between seminiferous tubules are **interstitial cells of Leydig,** which secrete **testosterone,** a male hormone
Epididymis	Convoluted seminiferous tubules become **straight tubules** that converge at **rete testes;** from testes tubules (ducts) become epididymis, where spermatozoa are temporarily stored for maturation; during ejaculation, spermatozoa are propelled by **peristaltic action**
Vas deferens	Continuous with epididymis; diameter increases, and ascends through inguinal canal into abdominal cavity; arches over side and down posterior surface of urinary bladder; during ejaculation, spermatazoa pass through vas deferens by peristaltic action
Ejaculatory duct	Continuous with vas deferens; located posterior to urinary bladder; here ducts from **seminal vesicles** unite and secrete alkaline viscous substance, which combines with sperm to form **semen**
Urethra	Joined by ejaculatory ducts of **prostate gland** and **bulbourethral gland;** these glands contribute to viscous substance of semen; duct at urinary bladder is closed during ejaculation
Penis	During sexual stimulation, arteries of penis dilate; this compresses veins so most entering blood is retained, erection occurs, providing effective copulatory organ; ejaculation may occur

KEY TERMS

Bulbourethral gland
Ejaculatory duct
Epididymis
Gametes
Germinal epithelium
Gonads
Interstitial cells of Leydig
Lobules
Male reproductive system
Peristaltic action
Penis
Primary spermatocytes
Prostate gland
Rete testes
Scrotum
Secondary spermatocytes
Semen
Seminal vessicles
Seminiferous tubules
Sertoli cells
Spermatids
Spermatogenesis
Spermatogonia
Spermatozoa
Sperm cells
Straight tubules
Testes
Testosterone
Tunica albuginea
Urethra
Vas deferens

FIG. 74-1 Male reproductive system (sagittal section).

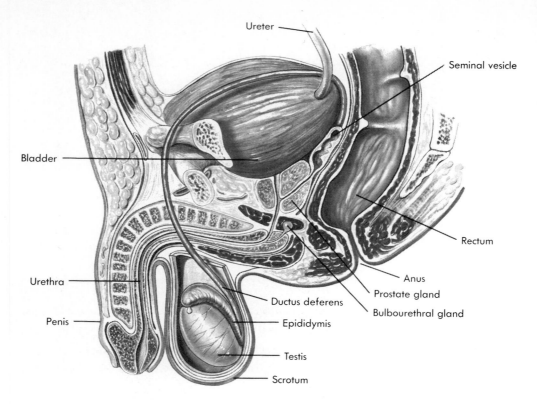

Ureter
Seminal vesicle
Bladder
Rectum
Anus
Prostate gland
Bulbourethral gland
Urethra
Ductus deferens
Penis
Epididymis
Testis
Scrotum

FIG. 74-2 **A,** Testes. **B,** Seminiferous tubule (cross-section).

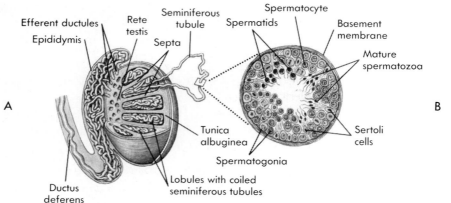

Efferent ductules
Rete testis
Seminiferous tubule
Spermatocyte
Spermatids
Basement membrane
Epididymis
Septa
Mature spermatozoa
A
B
Sertoli cells
Tunica albuginea
Ductus deferens
Lobules with coiled seminiferous tubules
Spermatogonia

PROCEDURE A   MALE REPRODUCTIVE GROSS ANATOMY

1. Locate the scrotum, then carefully remove it to expose the testes.
2. Locate the epididymis and trace it as it ascends to the vas deferens; the cat has no seminal vesicles.
3. Follow the vas deferens into the abdominal cavity to the prostate gland, where the duct passes into the urethra.
4. Cut the symphysis pubis, separating the pubic bone; follow the urethra downward toward the bulbourethral glands; at this point, the urethra enters the penis.
5. Explain the function of each structure.

PROCEDURE B   MALE REPRODUCTIVE HISTOLOGY

1. Place a prepared slide of the testis on low power of the microscope; observe the tunica albuginea and seminiferous tubules.
2. Turn to the high power objective and observe the various stages of sperm cells: basement membrane, spermatogonia, primary spermatocytes, secondary spermatocytes, and spermatids.
3. Identify associated Sertoli and interstitial cells of Leydig.
4. Draw and label your microscopic observations.

1. Explain the function of the scrotum.

2. Discuss the development of a sperm cell.

3. Draw and label the principal parts of a spermatozoan.

4. What is the function of the seminal vesicles, prostate gland, and bulbourethral gland?

5. Explain the function of testosterone.

*Use all references and materials at your disposal to answer these review questions.

75 REPRODUCTIVE SYSTEM
Female reproductive system

OBJECTIVES

1 List the structures associated with the ovaries.
2 Explain the development of the ovum.
3 Trace the course of an ovum through the female reproductive system.

MATERIALS

dissecting animal (female)　　　ovary prepared slide
dissecting equipment　　　　　　compound microscope

The **gonads,** or **ovaries,** of the **female reproductive system** function to produce **gametes,** or **ova,** and secrete hormones. **Ducts** of the reproductive system transport, receive, and store gametes. **Accessory glands** in the reproductive system secrete materials that support gametes. The female reproductive system structures (Fig. 75-1 and Color Plate 19) function as follows:

Ovaries	Attached to uterus by **ovarian ligament** and to pelvic wall by **suspensory ligament;** covering ovary is layer of simple cuboidal cells called **germinal (ovarian) epithelium** (Fig. 75-2); internal to this layer is capsule of collagenous connective tissue called **tunica albuginea** (see Fig. 75-2); internal to this are two regions of connective tissue called **stroma;** outer region is called **cortex,** and inner region is called the **medulla** (see Fig. 75-2); embedded in cortex are **ovarian follicles** containing **ova** in various stages of development; mature ovum and surrounding tissue is called **graffian follicle;** surrounding tissue secretes **estrogen;** after **ovulation,** release of ovum, follicular tissue becomes mass called **corpus luteum,** which secretes **progesterone;** corpus luteum eventually degenerates into **corpus albicans**
Uterine (fallopian) tubes	Fingerlike projections called **fimbriae** (see Fig. 75-2) lie close to ovary; ovum enters fimbriae and travels along the uterine tubes to **uterus** (see Fig. 75-2); peristaltic and ciliary action move ovum down into uterus; if ovum is fertilized by sperm, it occurs in upper third of uterine tube
Uterus	Highly muscular, inverted, pear-shaped sac with broad upper **fundus,** tapering to lower **cervix** (see Fig. 75-2); site of implantation of fertilized ovum and developing fetus; also site of **menstruation,** periodic discharge of epithelial cells, mucus, tissue fluid, and blood
Vagina	Receptacle for penis and passageway for menstrual flow; muscular tube with mucous membrane lining
Vulva	External genitalia

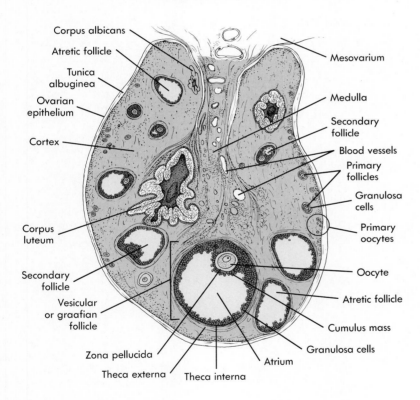

FIG. 75-1 Female reproductive structures.

Fundus
Ovary
Ovarian ligament
Fallopian tube
Round ligament
Body
Endometrium
Myometrium
Broad ligament (cut)
Cervix
Internal os
Cervical canal
Vagina (cut)
External os
Joan M. Beck

Corpus albicans
Atretic follicle
Tunica albuginea
Ovarian epithelium
Cortex
Mesovarium
Medulla
Secondary follicle
Blood vessels
Primary follicles
Granulosa cells
Primary oocytes
Corpus luteum
Secondary follicle
Vesicular or graafian follicle
Oocyte
Atretic follicle
Cumulus mass
Granulosa cells
Zona pellucida
Theca externa
Atrium
Theca interna

FIG. 75-2 Histology of ovary.

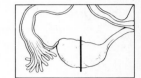

PROCEDURE A   FEMALE REPRODUCTIVE GROSS ANATOMY

1. Locate the ovaries just below the kidneys. Note the ligaments that support them.
2. Locate the uterine tubes and follow them to larger, thicker tubes called the uterine horn (uterus).
3. The uterine horns merge to form the body of uterus.
4. Cut the symphysis pubis, separating the pubic bone, then trace the uterus to the vagina.

PROCEDURE B   FEMALE REPRODUCTIVE HISTOLOGY

1. Place a prepared slide of the ovary on low power under the microscope. Observe the germinal epithelium and tunica albuginea.
2. Scan the field of view and observe the various stages of the maturing ovum.
3. Draw and label a cross-section of the ovary.

1. Explain how the ovaries are positioned in the pelvic cavity.

2. Describe the development of the ovum.

3. Compare the events that occur within the ovary to the menstrual cycle.

4. Define the following terms:

    Ectopic pregnancy
    Retroflexion of the uterus
    Anteflexion of the uterus
    Colposcopy
    Ovarian cysts
    Endometriosis

5. Discuss the cause of menopause.

*Use all references and materials at your disposal to answer these review questions.

REPRODUCTIVE SYSTEM
Meiosis

76

MATERIALS

Ascaris meiosis prepared slide   compound microscope

All human cells except gametes contain 23 pairs of homologous chromosomes—a **diploid** number (2n). Gametes contain one set of 23 chromosomes—a **haploid** number (n). **Reproduction** includes the process by which genetic material is passed from generation to generation. During this process, diploid cells divide into haploid cells (gametes) through special nuclear division called **meiosis** (Fig. 76-1).

In the **testes,** meiosis produces gametes called **spermatozoa** through the process called **spermatogenesis.** In the **ovary,** meiosis produces gametes called **ovum** through the process called **oogenesis.**

During **spermatogenesis** (Figs. 76-1 and 76-2) **spermatogonia,** diploid cells, undergo two stages of nuclear division. The deoxyribonucleic acid (DNA) is replicated in **interphase.** During **prophase I** and **metaphase I,** 23 homologous pairs (two **chromatids** each) line up along the **equatorial plane.** Each pair of homologous chromosomes (4 chromatids) form a **tetrad.** During this time, portions of a chromatid may exchange genetic information with another chromatid in a process called **crossing-over.** During this time the cells are known as **primary spermatocytes.** During **anaphase I,** pairs of chromosomes migrate to opposite poles, the first nuclear division is complete, and cells called **secondary spermatocytes** are formed. The cells are haploid, but the chromosomes are made up of two chromatids. (They are still duplicated since their centromere did not divide.) During **prophase II** and **metaphase II,** the 23 chromosomes (two chromatids) line up along the equatorial plane. The chromatids separate (centromere divides) from each other during **anaphase II,** forming **spermatids,** haploid cells containing half the number of chromosomes as the spermatogonia. Maturation of the spermatids forms **spermatozoa.** At the completion of spermatogenesis, four spermatozoa have developed from a single primary spermatocyte.

The diploid cell in **oogenesis** (Figs. 76-1 and 76-3) is called the **oogonium.** All reproductive cells complete the ability to undergo mitosis before birth of the female fetus. During **interphase I** the oogonium replicates DNA, forming tetrads, and crossing-over may occur. During this time the cells are known as **primary oocytes.** Once the primary oocyte is released and fertilized, it will undergo **prophase II** and **metaphase II.** During **anaphase II,** two cells of unequal size are produced. The larger cell receives most of the cytoplasm

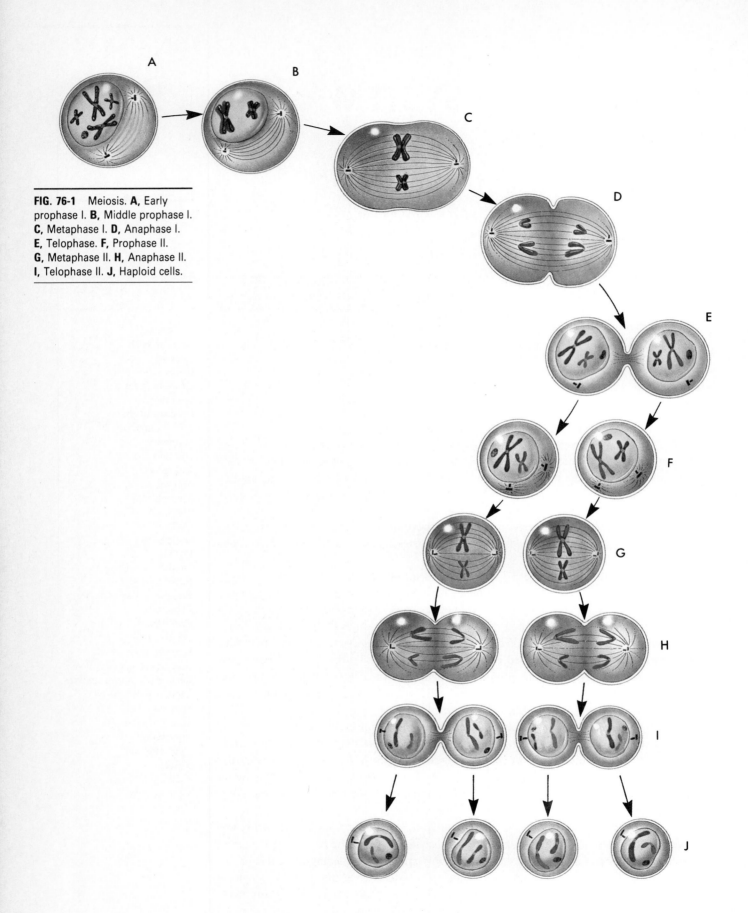

FIG. 76-1 Meiosis. **A,** Early prophase I. **B,** Middle prophase I. **C,** Metaphase I. **D,** Anaphase I. **E,** Telophase. **F,** Prophase II. **G,** Metaphase II. **H,** Anaphase II. **I,** Telophase II. **J,** Haploid cells.

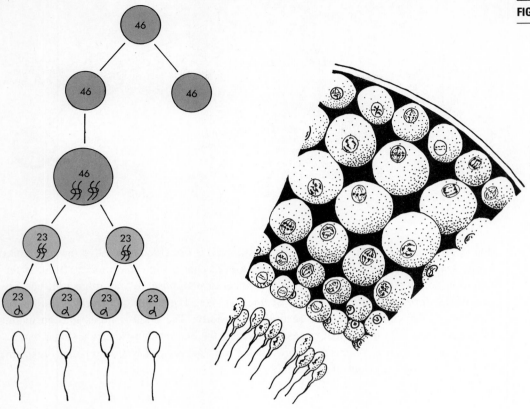

FIG. 76-2 Spermatogenesis.

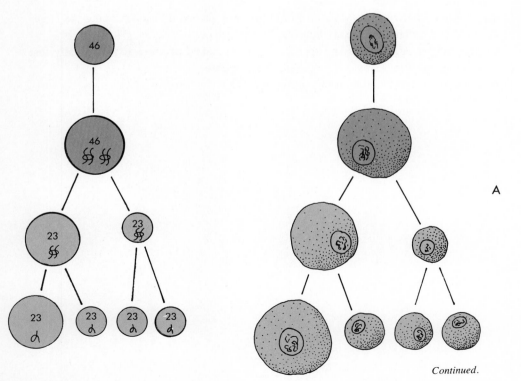

FIG. 76-3 **A,** Oogenesis.

A

Continued.

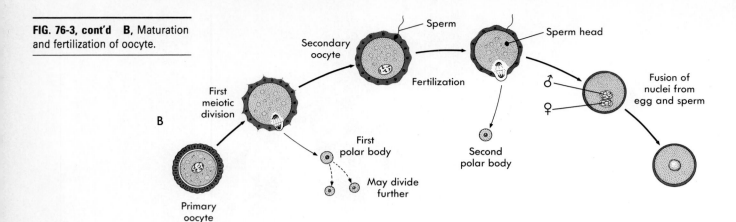

FIG. 76-3, cont'd B, Maturation and fertilization of oocyte.

Sperm

Secondary oocyte

Sperm head

First meiotic division

B

Fertilization

Fusion of nuclei from egg and sperm

♂
♀

First polar body

May divide further

Second polar body

Primary oocyte

and is called the **secondary oocyte.** The smaller cell is discarded and called the **first polar body.** During ovulation, the secondary oocyte is discharged into the **uterine tube.** If fertilization occurs, the secondary oocyte will undergo **prophase II, metaphase II,** and **anaphase II,** forming two haploid cells of unequal size. The larger cell is called the **ootid,** and the smaller cell is called the **second polar body.** The ootid matures into an ovum. Oogenesis is similar to spermatogenesis; however, the end result produces a single ovum.

Meiosis is easily observed in the roundworm, **Ascaris.** However, in this organism, meiosis does not occur until fertilization.

PROCEDURE A MEIOSIS IN *ASCARIS*

1. Focus the prepared slide on low power.
2. Locate the triangular shaped sperm outside the secondary oocyte.
3. Follow the sequence and note the large secondary oocyte with the small polar body attached to the surface.
4. Continue following the sequence, observing the second division forming the large ootid and second polar body attached to the surface.
5. The haploid egg nucleus and sperm nucleus are called **pronuclei.** They finally fuse together to form a single nucleus called the **zygote.**
6. Draw, label, and explain each state of meiosis.

1. Define the following terms:

   Diploid
   Haploid
   Meiosis
   Tetrad
   Crossing-over

2. Draw and label the process of spermatogenesis.

3. Draw and label the process of oogenesis.

4. Explain the processes associated with fertilization and implantation.

Use all references and materials at your disposal to answer these review questions.

APPENDIX

LABORATORY NOTEBOOK PROCEDURES

A hardbound laboratory notebook is recommended in order that the procedure and the results of an experiment may be maintained and organized. From the collected data one may easily review the experiment or write a laboratory report.

A suggested method for setting up a laboratory notebook is as follows:

	Lab title	Date
	INTRODUCTION	
	Paragraph describing purpose	
	of lab and terms necessary	
	to know	

The laboratory experiment title and date should be at the top of each page. The introduction should always begin on the right hand side of the notebook. If the introduction is more than one page, continue on the back side.

	Lab title	Date
MATERIALS		
List all necessary		
materials		
PROCEDURE	RESULTS	
List step by step how	State results simply	
you will perform the		
experiment		

Materials and procedure should always be on the left hand side of the notebook and the results of the right hand side, corresponding to the proper procedure.

WRITING A LABORATORY REPORT

A formal laboratory report should be designed similar to reports published in a scientific journal. The format is based on the scientific method. The purpose of the report is to state one's observations and hypotheses, to establish materials and procedures for testing the hypotheses, and to interpret results.

A format for writing a laboratory report is as follows:

Title

The title should reflect the purpose of the subject matter.

Introduction

The introduction should give a general background on information related to the investigation. The hypotheses and purpose of the experiment should be stated.

Materials

All materials necessary to perform the experiment should be listed.

Procedures

Steps of the procedures should give enough information so that someone else could repeat the experiment.

Results

Simply state, not discuss, the results obtained. Graphs, charts, tables, or diagrams often enhance the stated results.

Discussion

The discussion should be used to interpret and draw conclusions from the results obtained.

Literature cited

Cited literature should be on a separate page and should refer to any sources used when writing a lab report. Number and list all references alphabetically by author.

Example:

1. Adams, Robert E: Photosynthesis in Plants, pp. 155-156, St. Louis, 1985, The CV Mosby Co.

When using the reference in a report, only the number of the reference will be needed.

Example:

Plants use light as a source of energy for producing carbohydrates (1).

CREDITS

TEXT

8-2 William Ober

8-3, A Charles Flickenger: Medical cell biology, Philadelphia, 1979, WB Saunders Co.

8-3, B Ronald J Ervin

8-4, A Ronald J Ervin

8-4, B Richard Rodewald

8-5, A Charles Flickenger: Medical cell biology, Philadelphia, 1979, WB Saunders

8-5, B Ronald J Ervin

8-6 KG Murti, Visuals Unlimited

8-7 Ronald J. Ervin

8-8, A Kent McDonald

8-8, B Ronald J Ervin

8-9, A William Ober

8-9, C Susumo Ito, from Flickenger: Medical cell biology, Philadelphia, 1979, WB Saunders Co

15-1, 15-2 Michael Schenk

18-1 to 18-3 John V Hagen

19-1 Trent Stephens

24-1 to 24-5 David J Mascaro & Associates

25-1 to 25-5 David J Mascaro & Associates

26-1, 26-2 David J Mascaro & Associates

27-1 to 27-4 David J Mascaro & Associates

29-2 David J Mascaro & Associates

29-3 Rusty Jones

30-1 to 30-3 David J Mascaro & Associates

31-1 to 31-5 Terry Cockerham, Synapse Media Production

32-1 Rusty Jones

32-2 Janis Atlee and Michael Schenk

34 Review question illustration Joan Beck

42-1 Terry Cockerham

42-2 Michael Schenk

42-3 Michael Schenk

42-4, A Terry Cockerham

42-4, B Michael Schenk

43-1, 43-2, A to D Scott Bodell

47-1 Michael Schenk

47-2, A Trent Stephens

47-2, B Branislav Vidic

48-3 Branislav Vidic

48-4 Scott Bodell

50-1 Marsha Dohrmann

52-1 Marsha Dohrmann

53-1, A Marsha Dohrmann

53-1, B Kathy Mitchell Grey

54-1 Terry Cockerham

58-1, 58-2 Rusty Jones

58-3 Ronald J Ervin

61 Review question illlustration Rusty Jones

63-1 to 63-3 Ronald J Ervin

64-1 Joan Beck

64-2 Ronald J Ervin

64-2 Trent Stephens

65-1 Michael Schenk

66-1 Glanze, Walter D, editor: Mosby's medical and nursing dictionary, ed 2, St. Louis, 1986, The CV Mosby Co

66-2 Kathy Mitchell Grey

67-1 Jody Fulks

67-3 Thibodeau, Adapted in part from Lang and Wachsmith: Praktische Anatomic, Berlin, 1955, Springer-Verlag

67-4, A Jody Fulks

67-4, B Branislav Vidic

67-5 Thibodeau, Bevelander G, and Ramaley JA: Essentials of Histology, ed 8, St. Louis, The CV Mosby Co

68-1 Trent Stephens

68-2 Kaskel, Hammer, Kennedy, Oran: Laboratory biology, Merrill Publishing Co

69-3 David J Mascaro & Associates

69-4 Michael Schenk

70-1 Kathy Mitchell Grey

72-3, A David J Mascaro & Associates

72-3, B Branislav Vidic

72-3 inset Jody Fulks

74-1 Ronald J Ervin

74-2 William Ober

75-1 Joan Beck

75-2 Kevin Somerville

76-1 Ronald J Ervin

HISTOLOGY

Plate 1 Carolina Biological Supply Co

Plate 2, A to D Carolina Biological Supply Co

Plate 3, A and B Carolina Biological Supply Co

3, C Ed Reschke

Plate 4, A Ed Reschke

4, B and C Carolina Biological Supply Co

Plate 5, A Ed Reschke

5, B Carolina Biological Supply Co

Plate 6, A, B, D, and E Ed Reschke

6, C Trent Stephens

Plate 7, A to C Ed Reschke

Plate 8, A Trent Stephens

8, B Ed Reschke

Plate 9, A, C, and D Ed Reschke

9, B Carolina Biological Supply Co

Plate 10, A Carolina Biological Supply Co

10, B Ed Reschke

Plate 11, A and C Trent Stephens

11, B Carolina Biological Supply Co

Plate 12, A, B, and D Carolina Biological Supply Co

12, C Trent Stephens

12, E Ed Reschke

Plate 13 Pickup from Seeley, Stephens, Tate

Plate 14, A, C, D Ed Reschke

14, B Carolina Biological Supply Co.

Plate 15 Ed Reschke

Plate 16 Carolina Biological Supply Co

Plate 17, A, B and C Trent Stephens

Plate 18 Carolina Biological Supply Co

Plate 19 Carolina Biological Supply Co

INDEX

Base of microscope, 2
Basilar artery, 229
Basilar membrane, 203
Basilic vein, 233
Basophils, 212, 215
Benedict's reagent, 10-11
  in urine testing, 282
Beta cells, 209, 212
Biceps brachii muscle, 150
Biceps femoralis muscle, 156
Biceps reflex, 191, 192
Bicuspid valve, 220
Bile, 269
Bile duct, common, 269
Bile salts, 274
Binocular stereoscopic microscope, 3
Binocular vision and convergence, 200
Biological membranes, 18-22
Biological molecules, 10-18
  classes of, 10
Bipolar neuron, 169, 170
Bitter taste, 207
Bladder, urinary, 277
Blind spot test, 199
Blood, 214
  function of, 214
  histology of, 214-216
Blood cells
  classification of, 214-215
  red; see Erythrocytes
  white; see Leukocytes
Blood flow through heart, 220, 222
Blood groups, ABO, 248, 250
Blood pressure, 243, 246
  measurement of, 243, 244
Blood typing, 248-251
Blood vessels, 239
  histology of, 239-242
Blood-brain barrier, 170
Body
  ciliary, 194
  major arteries of, 229
  of mandible, 76
  nerve cell, 169
  planes of, 164, 166
  of rib, 83
  of sphenoid bone, 73
  of sternum, 83
  of stomach, 268
  of thoracic vertebra, 82
Body cavities, 164, 166
  classification of, 164
Body structure, planes of, 164
Bohr model of covalent bonding of water molecule, 6
Bolus, 268
Bond
  covalent, polar, 6, 7
  hydrogen, 7
  ionic, 7
  peptide, 15, 39
Bonding
  covalent
    of carbohydrates, 11
    of water molecule, 6
  hydrogen, of carbohydrates, 11
  ionic, of carbohydrates, 11
Bone, 63
  cancellous, 63, 64
  capitate, 90
  chemical properties of, 68-69

Bone—cont'd
  compact, 65
    histology of, 66
  conchae, inferior, 77
  cranial, 185
  ear, 77
  ethmoid, 74
  facial, 76
  frontal, 73
  hamate, 90
  hyoid, 80, 81
  lacrimal, 76
  long, 63
  lunate, 90
  mandible, 76
  maxilla, 77
  nasal, 76
  occipital, 74
  organic component of, 68
  palatine, 76
  pollex, 90
  processes of, 70
  pubic, 92
  scaphoid, 90
  of skull
    cranial, 73-75
    facial, 76-77
  sphenoid, 73-74
  spongy, 65
    histology of, 66
  temporal, 74
  trapezium, 90
  trapezoid, 90
  triquetrum pisiform, 90
  vomer, 77
  zygomatic, 76
Bone markings, 70-72
Bone marrow, 65, 214
Bony labyrinth, 203
Bowman's capsule, 277
Brachial artery, 229
Brachial plexus nerve, 172
Brachial region of nervous system, 172-174
Brachial vein, 233
Brachialis muscle, 151
Brachiocephalic vein, 233, 236
Brachioradialis muscle, 150
Brain, 185, 188
  gross anatomy of, 185-189
  midsagittal section of, 185, 187, 188
  ventricles of, 187, 188
Branched acinar exocrine glands, 50, 53
Bronchi, 256
Bronchial arteries, 229
Bronchioles, 256
  distribution of, 259
Brunner's glands, 268
Buccinator muscle, 137
Bulb
  end, Krause's, 124
  olfactory, 206
Bulbourethral gland, 285
Bundle, atrioventricular, 244
Bursae, 111
Bursitis, 111

C

Calcaneus of tarsal, 97
Calcium carbonate, 63, 68
Calcium phosphate, 63, 68

Callus, 101
Calyx, 277
Canal
  alimentary, 265, 272
  anal, 269
  central, of gray commissure, 181
  Haversian, 65
  of Schlemm, 194
  semicircular, 203
  Volkmann's, 65
Canaliculi, 65
Cancellous bone, 63, 64
Canthi, 193
Capacity, vital, 263
Capillaries, 239
  lymphatic, 252
  peritubular, 277
  sinusoid, of pituitary gland, 212
  structure of, 239
Capillary action, 7
Capitate bone, 90
Capitulum of humerus, 89
Capsule
  adipose, 277
  articular, 111
  Bowman's, 277
  fibrous, of articular capsule, 111
  of lymph nodes, 252
  renal, 277
Carbohydrate(s), 10-12
  categories of, 10
    formulas for, 10, 11
  digestion of, 274
Carboxyl group, 10, 11, 13, 16
Cardia portion of stomach, 268
Cardiac cycle, 220, 243-244
Cardiac muscle fibers, 224
Cardiac muscle tissue, 224-225
Cardiac sphincter; see Gastroesophageal sphincter
Cardiovascular system, 214-225, 228-251
Carotid arteries, 229
Carotid foramen, 78
Carpals, 90
Cartilage, 63, 105
  articular, 63
  corriculate, 255
  costal, of rib, 83
  cricoid, 255
  cuneiform, 255
  elastic, 60, 61
  hyaline, 60
  thyroid, 255
Cartilaginous articulations, 106
Cartilaginous connective tissue, 60-62
Cartilaginous joints, 105, 106, 108
Caruncle, lacrimal, 193
Catalytic proteins, 16
Cauda equina, 181
Caudal, 162
Caudofemoralis muscle, 156
Cavity(ies)
  abdominopelvic, 164, 167
    regions of, 165
  body, 164, 166
    classification of, 164
  glenoid, of scapula, 87
  medullary, 63
  pericardial, 164, 219
  pleural, 164, 356

Cecum, 269
Cell(s)
  air, mastoid, 74
  alpha, 209, 212
  animal, components of, 23
  beta, 209, 212
  blood; see Blood cells
  body, nerve, 169
  chief; see Zymogenic cells
  chromaffin, 209, 212
  epithelial
    squamous, 256
    stratified, 45
  glial, 170
  goblet, 49
  gustatory, 207
  haploid, 292
  hepatic, 268
  interstitial, of Leydig, 285
  metabolism of, 30-32
  mucous, 268
  nerve, 169
  olfactory, 206
  parafollicular, 209
  parietal, 268
  phagocytic, 252
  principal, 209, 212
    of parathyroid gland, 209
    of thyroid gland, 209, 212
  Schwann, 170
  septal, 256
  Sertoli, 285
  sperm, 285
  structure of, micrographs of, 23-29
  supporting
    of olfactory sense, 206
    of taste bud, 207
  thymus, 209, 253
  zymogenic, 268
Cell cycle, 35-37
Cell membrane, model, 19
Central canal of gray commissure, 181
Central fovea, 198
Centrioles, 26, 35
  electron micrograph of, 27
Centromere, 35
Cephalic, 162
Cephalic veins, 233
Cerebellum, 185
Cerebral aqueduct of brain, 188
Cerebral fissure, longitudinal, 185
Cerebral hemispheres, 185
Cerebrospinal fluid, 181
Cerebrum, 185, 188
Ceruminous glands, 202
Cervical enlargement of spinal cord, 181
Cervical region of nervous system, 172-174
Cervical vertebrae, 80
Cervix of uterus, 288
  of eye, 194
  of heart, 219
Chemical digestion, 274
Chemical groups, functional, 10, 11
Chief cells; see Zymogenic cells
Chloride ion, 7
Chlorides in urine, 282, 283
Chondrocytes, 60
Chordae tendineae, 220
Choroid, 194
Chromaffin cells, 209, 212

Chromatid, 35, 291
Chromatin, 35
Chromatogram, separation of amino acids using, 16
Chromosome, 33, 35
Chyme, 268
Cilia, 26
  electron micrographs of, 28
Ciliary body, 194
Circulation
  hepatic portal, 236
  systemic, 228
Circumduction, 117, 119
Circumvallate papillae, 206, 207
Cisternae, 25
Clavicle, 87, 88
Clavobrachialis muscle, 147
Clavodeltoid muscle, 147
Clavotrapezius muscle, 147
Clips, stage, of microscope, 2
Coarse-focus adjustment knob of microscope, 2
Coccyx, 83
Cochlea, 203
  fenestrated, 202
Cochlear branch, 203
Cochlear duct, 203
Codon combinations, 39
Codons, 38
Cohesion, 7
  and water, 8
Collagen, organic, 68
Collagenous connective tissue, 55, 57
Collagenous fibers, 55, 60, 63, 68
Collecting duct of kidney, 277
Colles' fracture, 101, 102
Colloid, thyroid, 209, 212
Colon, 269
Column
  spinal, 80
  vertebral, 80
Columnar epithelium
  simple, 41, 43
  stratified, 45, 47
Comminuted fracture, 101, 102
Commissure, gray, 181
Common bile duct, 269
Common carotid arteries, 229
Common hepatic artery, 229
Common hepatic duct, 269
Common iliac arteries, 229
Common iliac vein, 237
Common peroneal and tibial nerve, 178
Compact bone, 65
  histology of, 66
Complementary pair, 33
Complete fracture, 101
Compound acinar exocrine glands, 50, 53
Compound fracture, 101, 102
Compound light microscope, 2
Compound microscope observation of skeletal muscle tissue, 132
Compound tubular exocrine glands, 50, 51
Concentric rings, 65
Conchae, 74
Conchae bones, inferior, 77
Condensation reaction, 11
Condenser of microscope, 2
Conduction deafness, 203
Condyles, 70, 74
  of femur, 96

Condyles—cont'd
  of mandible, 76
  of tibia, 97
Cones, 194
Conjunctiva, 194
Connective tissue, 55-62
  adipose, 55, 56
  adult, 55-59
  cartilaginous, 60-62
  dense, 55, 57
  elastic, 55, 57
  embryonic, 55
  fibrous, 105
  loose, 55, 56
  reticular, 55, 58
Constriction of iris muscles, 198
Contractile proteins, 16
Conus medullaris, 181
Convergence, 198
  binocular vision and, 200
Convoluted tubule, 276
Coracoid process of scapula, 87
Cornea, 194
Coronal plane, 164
Coronal suture of frontal bone, 73
Coronary arteries, 229
Coronary sinus, 220
Coronoid fossa of humerus, 89
Coronoid process
  of mandible, 76
  of ulna, 89
Corpus albicans, 288
Corpus callosum, 185
Corpus luteum, 288
Corpuscles, 124
Corriculate cartilage, 255
Corrugator supercilii muscle, 137
Cortex
  adrenal, 209, 212
  of lymph node, 252
  of spleen, 252
Cortex region of ovary, 288
Corti, organ of, 203
Cortical nephrons, 277
Costal cartilage of rib, 83
Covalent bonding
  of carbohydrates, 11
  in DNA, 33
  polar, 6, 7
  of water molecule, 6
Coxal joint, 111, 113, 114
Cranial, 162
Cranial bones, 73-75, 185
Cranial cavity, 164
Cranial meninges, 185
Cranial nerves, 172, 185, 186
  identification of, 188
Cranial nerve X, 172
Cranial nerve XI, 172
Cranial nerve XII, 172
Crenation, 217
Crest (of bone), 70
  iliac, 92
  pubic, 94
  of tibia, 97
Cribriform plate, 74, 206
Cricoid cartilage, 255
Crista galli, 74
Cristae, 26
Crossing-over, 291
Crypts of Lieberkühn, 268

Gluteus medius muscle, 156
Glycerol, 13
Glycolipids, 18
Glycolysis, 30
Glycoproteins, 18
Goblet cell, 49
Golgi complex, 26
Gomphosis, 105
Gonadal arteries, 229
Gonads, 285, 288
Gracilis muscle, 155
Graffian follicle, 288
Granulocytes, 214
Gray commissure, 181
Gray horns, 181
Gray matter, 181
Greater curvature of stomach, 268
Greater omentum, 269
Greater palatine foramen, 78
Greater sciatic notch of ilium, 92
Greater trochanter of femur, 96
Greater tubercle of humerus, 87
Greater wings of sphenoid bone, 73
Greenstick fracture, 101, 102
Groove (of bone), 70
    lacrimal, of maxilla, 77
    radial, of humerus, 89
Growth hormone, human, 212
Guanine, 33, 38
Gustatory cells, 207
Gustatory hair, 207
Gustatory sensations, 206-207
Gyri of brain, 185

H

Hair, 123
    gustatory, 207
    olfactory, 206
Hair end organs, 124
Hairpin loop of Henle, 277
Hamate bone, 90
Hammer; *see* Malleus
Hamstring muscles; *see* Biceps femoralis mus-
    cle; Semimembranosus muscle; Semi-
    tendinosus muscle
Haploid cells, 292
Haploid number of chromosomes, 291
Haversian canals, 65
Head
    of bone, 70
    of femur, 96
    of humerus, 87
    muscles of, 142
    and neck
        muscles of, 140, 142, 143
        veins of, 233
    of radius, 90
    of rib, 83
    of ulna, 89
Hearing, 202-205
Heart, 219
    anatomy of, 219-223
    blood flow through, 220, 222
    chambers of, 219
    histology of, 220
Heart wall, layers of, 224
Heartbeat, 220
Heat, specific, 7
Hematoma, fracture, 101
Hematopoiesis, 214

Hemiazygous vein, 236
Hemispheres, cerebral, 185, 188
Hemoglobin, 214
Hemolysis, 217
Henle, loop of, 277
Hepatic artery, common, 229
Hepatic cells, 269
Hepatic duct, common, 269
Hepatic portal circulation, 237
Hepatic vein, 236
HGH; *see* Human growth hormone
Hilum
    of kidney, 277
    of lymph nodes, 252
    of spleen, 252
Hindlimb region, muscles of, 154-159
Hinge joint, 106, 107
Hip joint, 111, 113, 114
Histones, 33, 35
Holocrine gland, 49
Horizontal plane, 165
Horizontal plate, 76
Hormone(s), 209, 212
    human growth, 212
    parathyroid, 209
Human growth hormone, 212
Human reflexes, 190-192
Humerus, 87, 89
Humor
    aqueous, 194
    vitreous, 194
Hyaline cartilage, 60
Hydrochloric acid, 274
Hydrogen bond, 7
Hydrogen bonding of carbohydrates, 11
Hydrolysis, 11
    of proteins, 16
Hydrophilic molecules, 7
Hydrophobic molecules, 7
Hydroxyapatite, 68
Hydroxyl group, 10, 11, 13
Hyoid bone, 80, 81
Hyperextension, 117
Hypertonic environment, 18, 19
Hypertonic solution, 19, 217
Hypoglossal foramen, 78
Hypoglossal nerve, 172, 185
Hypothalamus, 185, 209
Hypotonic environment, 18, 19
Hypotonic solution, 19, 217

I

Ileocecal valve, 269
Ileum, 268, 269
Iliac arteries, 229
Iliac crests, 92
Iliac fossa, 92
Iliac veins, 236
Iliacus muscle; *see* Iliopsoas muscle
Iliocostalis muscle, 147
Iliopectineal line of ilium, 92
Iliopsoas muscle, 155
Ilium, 92
Immunological proteins, 16
Impacted fracture, 101, 102
Impulses
    motor, 181
    sensory, 181
Incisive foramen, 78
Incomplete fracture, 101, 102

Incus, 77, 202
Inferior, 162
Inferior articulating process of thoracic verte-
    bra, 82
Inferior conchae bones, 77
Inferior mesenteric arteries, 229
Inferior nuchal line, 74
Inferior phrenic arteries, 229
Inferior phrenic vein, 236
Inferior pubic ramus, 94
Inferior sagittal sinus vein, 233
Inferior spines of ilium, 92
Inferior vena cava, 220, 236
Infraspinatus muscle, 147
Infundibulum, 209
Inlet, pubic, 94
Inner ear, 202, 203, 204
Inner layer
    of alimentary canal; *see* Mucosa layer of ali-
        mentary canal
    of periosteum, 63
Inorganic salts, 68
Insertion of muscle tendon, 140, 145, 149
Inspiration, 263
Inspiratory reserve volume, 263
Insulin, 209, 212, 268
Integrated proteins, 18
Integumentary system, 123-128
    histology of, 123-126
Intercalated disks of cardiac muscle fibers, 224
Intercellular material; *see* Matrix
Intercondylar eminence of tibia, 97
Intercondyloid notch of femur, 96
Intercostal arteries, 229
Intercostal muscles, 142
Internal auditory meatus, 74
Internal carotid artery, 229
Internal iliac arteries, 229
Internal intercostal muscles, 142
Internal jugular veins, 233
Internal oblique muscle, 142
Internal occipital protuberance, 74
Internal respiration, 255
Interphase, 35, 291
Interphase I, 291
Interstitial cells of Leydig, 285
Interstitial fluid, 214
Intertubercular of humerus, 87
Intestinal glands, 265
Intestine
    large, 265, 269
    small, 265, 268
Intracellular membranes, 18
Intraocular pressure, 194
Intraorbital foramen, 78
Intraventricular foramen of brain, 188
Inversion, 117, 120
Invertor muscle, 134
Iodine reagent test, starch and, 11
Ion
    chloride, 7
    sodium, 7
Ionic bond, 7
Ionic bonding of carbohydrates, 11
Iris, 194
    constriction of, muscles of, 198
Iris diaphragm of microscope, 2
Iron in hemoglobin, 214
Ischial spine, 94
Ischial tuberosity, 92

Ischium, 92
Islets of Langerhans, 209, 212, 268
Isotonic environment, 18
Isotonic solution, 19, 217

J

Jejunum, 267
Joint, 105-108; *see also* Articulation(s)
  ball-and-socket, 107
  cartilaginous, 105, 106, 108
  classification of, 105-108
  coxal, 111, 113, 114
  diarthrotic, movement of, 117-122
  ellipsoidal, 107
  fibrous, 105, 108
  gliding, 105, 106, 107
  hinge, 106, 107
  hip, 111, 113, 114
  knee, 111-112, 114
  pivot, 106, 107
  saddle, 107
  synovial, 105, 106-107, 108, 110-116
    structure of, 110
  tibiofemoral, 111-112, 114
Jugular foramen, 78
Jugular fossa, 74
Jugular vein, 233
Juxtamedullary nephrons, 277

K

Keratin, 123
Keratohyalin, 123
Ketone group, 10, 11
Kidneys, 277, 279, 280
Kinesiology, 160-161
Knee jerk, 190, 191, 192
Knee joint, 111-112, 114
Knobs of microscope, adjustment, 2
Krause's end bulb, 124
Krebs cycle, 30

L

Labyrinth, 203
Lacerum foramen, 78
Lacrimal bones, 76
Lacrimal caruncle, 193
Lacrimal duct, 194
Lacrimal foramen, 78
Lacrimal gland, 194
Lacrimal groove of maxilla, 77
Lacrimal secretions, 194
Lacunae, 65
Lambdoidal suture, 74
Lamellae, 65
Lamina propria, 271
Laminae of thoracic vertebra, 82
Langerhans, islets of, 209, 212
Large intestine, 265, 269
Larynx, 255
Lateral, 162
Lateral apertures of brain, 188
Lateral arches of foot, 99
Lateral canthi, 193
Lateral cutaneous nerve, 178
Lateral epicondyle of humerus, 89
Lateral malleolus of fibula, 97
Lateral muscles
  of forelimb, 151
  of hindlimb, 156
Lateral rotation, 117, 119

Lateral ventricles, 188
Lateral white column, 182
Latissimus dorsi muscle, 142, 147
Lattice, microtubrical, of cytoskeleton, 24, 25
Lectins, 248-249
  and blood agglutination procedures, 250
Left atrium, 219
Left hemisphere of brain, 188
Left ventricle, 219
Lens, 194
Lens accommodation near point, 199
Lesser omentum, 269
Lesser palatine foramen, 78
Lesser trochanter of femur, 96
Lesser tubercle of humerus, 87
Lesser wings of sphenoid bone, 73
Leukocytes, 214
  identification of, 215
Levator muscle, 134
Levator scapulae ventralis muscle, 147
Leydig, interstitial cells of, 285
Lieberkühn, crypts of, 268
Ligament(s), 111
  falciform, 269
  ovarian, 288
  suspensory, 288
    of lens, 194
Limb of Henle, 277
Line of bone, 70
  epiphyseal, 63
  gluteral, of ilium, 92
  iliopectineal, 92
  nuchal, 74
  popliteal, of tibia, 97
Linea aspera of femur, 96
Lingual tonsils, 252
Lipases, 274
Lipid(s), 13-14
Lipid tail of plasma membrane, 18
Litmus, 274
Liver, 265, 269
Lobules of tunica albuginea, 285
Long bone
  anatomy of, 63-65
  histology of, 66
Long thoracic nerve, 173
Longissimus capitis muscle, 147
Longissimus cervicis muscle, 147
Longissimus dorsi muscle, 147
Longitudinal arches of foot, 99
Longitudinal cerebral fissure, 185
Loop of Henle, 277
Loose connective tissue, 56
Lower extremities, 96-100
  veins of, 236
Lower and upper eyelids, junctions of, 193
Lumbar arteries, 229
Lumbar enlargement of spinal cord, 181
Lumbar region
  muscles of, 145, 147
  of nervous system, 178-180
Lumbar veins, 236
Lumbar vertebrae, 83
Lumbosacral nerve cord, 178
Lumbosacral plexus, 178
Lumen of blood vessels, 239
Lunate bones, 90
Lung capacity, 262-264
Lung volumes, 262
Lungs, 256, 258

Lymph, 252
Lymph nodes, 214, 252, 253
Lymph vessels, 252
Lymphatic capillaries, 252
Lymphatic system, 252-254
Lymphocytes, 215
Lysosomes, 26
Lysozyme, 194, 265

M

Magnichanger knob of microscope, 2
Magnum foramen, 78
Major calyx, 277
Male reproductive system, 285-287
Malleolus
  lateral, of fibula, 97
  medial, of tibia, 97
Malleus, 77, 202
Mandible, 76
Mandibular foramen, 78
Mandibular fossa, 74
Manubrium of sternum, 83
Markings, bone, 70-72
Marrow, bone, 65
Masseter muscle, 142
Mastication, 265
Mastoid air cells, 74
Mastoid foramen, 79
Mastoid process, 74
Matrix, 55, 63
Maxilla, 77
Maxillary sinus, 77
Meatus (of bone), 70
  auditory, 74, 202
Medial, 162
Medial arches of foot, 99
Medial canthi, 193
Medial cubital vein, 233
Medial epicondyle of humerus, 89
Medial malleolus of tibia, 97
Medial muscles
  of forelimb, 150
  of hindlimb, 155-156
Medial rotation, 117, 119
Median nerve, 173
Medulla
  adrenal, 209, 212
  of lymph node, 252
  of spleen, 252
Medulla oblongata, 185
Medulla region of ovary, 288
Medullary cavity, 63
Medullary sinuses, 252
Megakaryocytes, 215
Meibomian glands, 193
Meiosis, 291-295
Meissner, plexus of, 272
Meissner's corpuscles, 124
Membranes
  biological, 18-22
  cell, model, 19
  intracellular, 18
  nuclear, 35
  plasma; *see* Plasma membrane
  pleural, 256, 260
  serous, 269
  synovial, 111
  tympanic, 202
Membranous labyrinth, 203
Meninges, 181

Parotid glands, 265
Partial fracture, 101, 102
Patella, 96
Patellar reflex, 190, 191, 192
Pectineus muscle, 155
Pectoantebrachialis muscle, 142
Pectoral girdle, 87
Pectoralis major muscle, 142
Pectoralis minor muscle, 142
Pedicles of thoracic vertebra, 82
Pelvic cavity, 164
Pelvic girdle, 92-95
Pelvic outlet, 94
Pelvis, 93
    abdomen and, veins of, 236
    false, 94
    male and female, comparison of, 94,
        95
    renal, 277
    true, 94
Penis, 284
Pepsin, 273
Pepsinogen, 268, 274
Peptide, 15
Peptide bonds, 15, 39
Pericardial cavity, 164, 219
Pericardial fluid, 219
Pericardial sac, 219, 222
Pericardium, 219
Perilymph, 203
Perimysium, 129
Periosteum, 63
Peristaltic action, 268
Peritoneum, 269
Peritubular capillaries, 277
Peroneal and tibial nerve, common, 178
Peroneal vein, 236
Perpendicular plate, 74
Petrous portion of temporal bone, 74
Phagocytic cells, 252
Phalanges, 90, 97
    of foot, 97, 98
    of hand, 90
Pharyngeal tonsils, 252
Pharynx, 255, 265, 268
Phosphate, 33, 38
    in DNA, 33
    of RNA, 38
    in urine, 282, 283
Phosphate group, 10, 11
    of plasma membrane, 18
    polar, 13
Phospholipid bilayer of plasma membrane, 18,
    23
Phospholipids, 13-14
Phosphorylation, oxidative, 30
Photoreceptors, 198
Phrenic arteries, 229
Phrenic nerve, 173
Phrenic vein, 236
Pia mater, 181
Pigmented layer of retina, 194
Pineal gland, 209
Pinna; see Auricle
Pituitary gland, 209, 212
Pivot joint, 106, 107
Plane(s), 164, 166, 167
    body, 164, 166
    equatorial, 291
    organ, 166

Plantar extension, 117
Plantar flexion, 119
Plantar flexor muscle, 134
Plantar reflex, 191, 192
Plasma, 214
Plasma membrane, 18, 23
    of cardiac muscle fibers, 224
    electron micrograph of, 23
    of muscle fiber, 129
Plate
    cribriform, 74, 206
    epiphyseal, 63
    horizontal, 76
    perpendicular, 74
    tarsal, 193
Platelets; see Thrombocytes
Platysma muscle, 137
Pleura, 256
Pleural cavities, 164, 356
Pleural membrane, 256, 260
Plexus
    of Auerbach, 272
    lumbosacral, 178
    of Meissner, 272
Polar bodies, 294
Polar covalent bond, 6, 7
Polar phosphate group, 13
Pollex bones, 90
Polypeptide, 15
Polysaccharides, 11
Pons varolii, 185
Popliteal line of tibia, 97
Popliteal vein, 236
Pore
    nuclear, 24, 25
    taste, 207
Portal circulation, hepatic, 236
Position, anatomical, 117, 118, 163
Posterior, 162
Posterior chamber of eye, 194
Posterior gray horns, 181
Posterior inferior spine of ilium, 92
Posterior superior spine of ilium, 92
Posterior white columns, 182
Pott's fracture, 101, 102
Pressure
    blood; see Blood pressure
    intraocular, 194
Pressure stimuli, 124
Primary bronchi, 256
Primary oocyte, 291
Primary spermatocytes, 285, 291
Prime movers, 160
Principal cells
    of parathyroid gland, 209
    of thyroid gland, 209, 212
Procerus muscle, 137
Process(es) (of bone), 70
    acromion, of scapula, 87
    alveolar
        of mandible, 76
        of maxilla, 77
    articulating, of thoracic vertebra, 82
    coracoid, of scapula, 87
    coronoid
        of mandible, 76
        of ulna, 89
    mastoid, 74
    olecranon, of ulna, 89

Process(es)—cont'd
    pterygoid, 74
    spinous, 70
        of thoracic vertebra, 82
    styloid, 74
        of radius, 90
        of ulna, 89
    transverse, of thoracic vertebra, 82
    zygomatic, 74
Progesterone, 287
Pronation, 117, 120
Pronator muscle, 134
Pronator teres muscle, 150
Prophase, 35
Prophase I, 291, 292
Prophase II, 291, 292, 294
Prostate gland, 285
Proteins, 15-17
    catalytic, 16
    contractile, 16
    detection of, 16
    digestion of, 275
    immunological, 16
    integrated, 18
    regulator, 16
    structural, 16
    surface, 18
    transport, 16
Protraction, 117, 120
Proximal, 162
Proximal convoluted tubule, 277
Pseudostratified epithelium, 45, 47
Psoas major muscle; see Iliopsoas muscle
Pterygoid processes, 74
Ptyalin, 274
Pubic arch, 94
Pubic bone, 92
Pubic crest, 94
Pubic inlet, 94
Pubic ramus, 94
Pubic tubercle, 94
Pubis; see Pubic bone
Pulmonary semilunar valve, 220
Pulmonary veins, 220
Pulp
    red, 253
    white, 252
Pulse, 243
Pulse rate, 246
Punctum, 194
Pupil, 194
Pupillary reflexes, 200
Purkinje fibers, 244
Pyloric sphincter, 268
Pylorus, 268
Pyruvate, 30

Q

QRS wave, 43

R

Radial artery, 229
Radial groove of humerus, 89
Radial nerve, 173
Radial tuberosity, 90
Radioulnar articulation, 106
Radius, 90
Ramus
    of mandible, 76
    pubic, 94
Reaction, condensation, 11

Sphincter
  cardiac, 268
  gastroesophageal, 268
  of Oddi, 269
  pyloric, 268
Sphincter muscle, 134
Sphygmomanometer, 243, 244
Spinal accessory nerve, 172
Spinal column, 80, 81
Spinal cord, 181
  cross section of, 183
  damage to, and reflexes, 191
  gross anatomy of, 181-184
  and spinal nerves, 182
Spinal foramen of thoracic vertebra, 82
Spinal ganglia, 181, 183
Spinal nerves, 172
  and spinal cord, 182
Spinal trunk, thoracic, 175
Spinalis dorsi muscle, 147
Spindle, mitotic, 35
Spine
  of ilium, 92
  of ischium, 94
  of scapula, 87
Spinosum foramen, 79
Spinotrapezius muscle, 147
Spinous process
  of bone, 70
  of thoracic vertebra, 82
Spiral fracture, 101, 102
Spirometer, 263
Spleen, 214, 252, 253
Splenic artery, 229
Splenius muscle, 147
Spongy bone, 65
  histology of, 66
Squamosal suture, 74
Squamous epithelial cells, 256
Squamous epithelium
  simple, 41, 42
  stratified, 45, 46
Squamous portion of temporal bone, 74
Stability, temperature, and water, 8
Stage clips of microscope, 2
Stages of microscope, 2
Stapes, 76, 202
Starch, 11
  and iodine reagent test, 11
Stensen's duct, 265
Sternocleidomastoid muscle, 142
Sternohyoid muscle, 142
Sternum, 80, 83, 84, 85
Stethoscope and blood pressure measurement,
    243, 244
Stimuli, 124
Stirrup; see Stapes
Stomach, 265, 268
Straight sinus vein, 233
Straight tubules, 254
Strand
  non-transcribed, 38
  transcribing, 38
Stratified columnar epithelium, 45, 47
Stratified cuboidal epithelium, 45, 46
Stratified epithelial cells, 45
Stratified epithelium, 45-48
Stratified squamous epithelium, 45, 46
Stratum basale, 123
Stratum corneum, 123
Stratum granulosum, 123

Stratum lucidum, 123
Stratum spinosum, 123
Striated muscle tissue, 129
Stroma region of ovary, 288
Structural model of covalent bonding of water
    molecule, 6
Structural proteins, 16
Structure, body, planes of, 164
Styloid process, 74
  of radius, 90
  of ulna, 89
Stylomastoid foramen, 79
Subclavian artery, 229
Subclavian vein, 233, 236
Subgroups in blood typing, 248
Sublingual glands, 268
Submandibular glands, 268
Submucosa layer of alimentary canal, 272
Subscapular nerve, 172
Subscapularis muscle, 142
Sucrose, 11
Sudan III reagent in lipid identification, 13, 14
"Suicide bag," 26
Sulcus
  of bone, 70
  of brain, 185
Superciliary arches of frontal bone, 73
Superficial, 162
Superficial muscles
  of abdominal region, 142
  of forelimb, 150, 151
  of hindlimb, 155, 156
  of neck and head, 142
  of shoulder and thorax, 147
  of thorax, 142, 147
Superior, 162
Superior articulating process of thoracic verte-
    bra, 82
Superior border of scapula, 87
Superior mesenteric artery, 229
Superior nuchal line, 74
Superior orbital fissure, 74
Superior phrenic arteries, 229
Superior pubic ramus, 94
Superior sagittal sinus vein, 233
Superior spine of ilium, 92
Superior turbinates, 74
Superior vena cava, 220, 236
Supination, 117, 120
Supinator muscle, 134, 151
Supporting cells
  of olfactory sense, 206
  of taste bud, 207
Supracondylar ridges of femur, 96
Supraorbital foramen, 79
Supraorbital margin of frontal bone, 73
Suprarenal artery, 229
Suprascapular nerve, 172
Supraspinatus muscle, 147
Sural nerve, 178
Surface proteins, 18
Surface tension, 7
  of water, 8
Surgical neck of humerus, 87
Suspensory ligament, 288
  of lens, 194
Suture, 105
  coronal, 73
  lambdoidal, 74
  sagittal, 73
  squamosal, 74

Sweet taste, 207
Sympathetic nerve trunks, 175
Sympathetic trunk nerve, 172, 175
Symphysis, 106
Symphysis pubis, 94
Synarthroses, 105
Synchondrosis, 106
Syndesmosis, 105, 106
Synergists, 160
Synovial articulation, 106-108
Synovial joints, 105, 106-107, 108, 110-116
  structure of, 110
Synovial membrane, 111
Synthesis, dehydration; see Dehydration syn-
    thesis
Synthesis phase; see S phase
Systemic circulation, 228
Systemic veins, 233
Systole, 220, 243
Systolic blood pressure, 243

T

T wave, 244
T$_3$; see Triiodothyronine
T$_4$; see Thyroxin
Tactility, 124
Talus of tarsal, 97
Tarsal plate, 193
Tarsals, 97, 98
Taste, sense of, 206-207
Taste buds, 206-207
Taste pore, 207
Taste test, 207
Tastes, 207
T-cells, 209, 253
Teeth, 265
Telophase, 35, 36
Temperature stability and water, 8
Temperature stimuli, 124
Temporal bones, 74
Temporalis muscle, 142
Tendon, 129
  muscle, 140, 145, 149
  reflexes of, 191
Tension, surface, 7
  of water, 8
Tensor fasciae latae muscle, 156
Tensor muscle, 134
Teres major muscle, 147
Teres minor muscle, 147
Terminal bronchioles, 256
Terms, directional, 162
Tertiary bronchi, 256
Test
  afterimage, 200
  blind spot, 199
  iodine reagent, starch and, 11
  monosaccharide reagent, 11
  taste, 207
  urine, 282-283
Testes, 284, 286
  meiosis in, 291
Testicular arteries, 229
Testicular vein, 236
Testosterone, 285
Tetrad, 291
Thalmus, 185
Third ventricle, 188
Thoracic aorta, 228, 229
Thoracic cavity, 164
Thoracic nerve, 173

Thoracic region of nervous system, 175-177
Thoracic spinal trunk nerves, 175
Thoracic vertebrae, 82
Thorax, 83
    muscles of, 140, 142, 143, 145, 146, 147
    veins of, 236
Throat; *see* Pharynx
Thrombocytes, 214, 215
    identification of, 215
Thymine, 33
Thymus, 252
Thymus cells, 209, 253
Thymus gland, 209, 253
Thyroid cartilage, 255
Thyroid colloid, 209, 212
Thyroid follicle, 212
Thyroid gland, 209, 212
Thyroxin, 209, 212
Tibia, 97, 98
Tibial nerve, peroneal and, common, 178
Tibial tuberosity, 97
Tibial vein, 236
Tibialis anterior muscle, 155
Tibiofemoral joint, 111-112, 114
Tibiofibular articulation, 106
Tidal volume, 263
Tissue
    connective; *see* Connective tissue
    epithelial, 41-54; *see also* Epithelial tissue
    muscle; *see* Muscle tissue
    osseous, 63
    serous, of heart, 224
Tongue, 265
    microscopic observation of, 207
    papillae of, 206, 207
Tonsils, 214, 252
Trabeculae, 65
Trabeculum
    of lymph node, 252
    of spleen, 252
Trachea, 255, 258
Tracts of white columns, 182
Transcribing strand, 38
Transcription, DNA, 38-40
Transfer RNA, 38
Translation, DNA, 38-40
Transmission electron microscope, 23
Transport proteins, 16
Transverse arches of foot, 99
Transverse fracture, 101
Transverse plane, 164
Transverse process of thoracic vertebra, 82
Transverse sinus vein, 233
Transversus abdominis muscle, 142
Transversuscostarum costeria muscle, 142
Trapezium bone, 90
Trapezoid bone, 90
Triceps brachii muscle, 150
Triceps reflex, 191, 192
Tricuspid valve, 220
Trigeminal nerve, 185
Triglyceride, 13
Triiodothyronine, 209, 212
Triquetrum pisiform bone, 90
tRNA; *see* Transfer RNA
Trochanter, 70
    of femur, 96
Trochlea
    of femur, 96
    of humerus, 89
Trochlear nerve, 185

True pelvis, 94
True ribs, 83
Trunks
    sympathetic nerve, 172, 175
    thoracic spinal, 175
Tube
    auditory, 202
    eustachian, 202
    fallopian, 288, 294
    uterine, 288, 294
Tubercle (of bone), 70
    gluteal, of femur, 96
    of humerus, 87
    pubic, 94
    of rib, 83
Tuberosity (of bone), 70
    deltoid, of humerus, 89
    frontal, 73
    gluteal, of femur, 96
    ischial, 92
    radial, 90
    tibial, 97
Tubular exocrine gland
    compound, 50, 51
    simple coiled, 50, 52
    simple straight, 50, 51
Tubule
    convoluted, 277
    seminiferous, 285, 286
    straight, 285
Tubuloacinar gland, 194, 265
Tunica albuginea, 285, 288
Tunica externa, 239
Tunica intima, 239
Tunica media, 239
Tunica mucosa, 272
Tunica muscularis, 272
Tunica serosa, 272
Tunica submucosa, 272
Turbinates, 74
Tympanic membrane, 202
Type A blood, 248, 249
Type AB blood, 248, 249
Type B blood, 248, 249
Type O blood, 248, 249

U

Ulna, 89, 90
Ulnar artery, 229
Ulnar nerve, 173
Unicellular exocrine glands, 50
Unicellular glands, 49
Unipolar neuron, 169, 170
Upper extremities, 87-91
    veins of, 233
Upper and lower eyelids, junctions of, 193
Uracil, 38
Urea, 282
Ureters, 277, 279, 280
Urethra, 277, 285
Urinary bladder, 277
Urinary system, 277-284
    gross anatomy and histology of, 277-281
    physiology of, 282-284
Urine, 282
Urine testing, 282-283
Urogenital system, 278
Uterine tube, 288, 294
Uterus, 285
Utricle, 203

V

Vagina, 285
Vagus nerve, 172, 175, 185
    and taste impulses, 207
Valves
    of heart, 219-220
    ileocecal, 269
    in veins, 238, 241
Vas deferens, 285
Vasa recta, 277
Vascular tunic, 194
Vastus intermedius muscle, 156
Vastus lateralis muscle, 155
Vastus medialis muscle, 156
Vater, ampulla of, 269
Vein(s), 233-237, 239
    of abdomen and pelvis, 236
    of head and neck, 233
    of lower extremities, 236
    microscopic observations of, 242
    pulmonary, 220
    renal, 236, 277
    of upper extremities, 233
    valves in, 239, 241
    walls of, 239, 240
Vena cava, 220, 236
Ventilation, 255
Ventral, 162
Ventral cavity, 164
Ventral gray horns, 181
Ventral root of spinal cord, 181
Ventral thoracic nerve, 173
Ventricles, 188
    of brain, 187, 188
    of heart, 219
Venules, 238
    renal, 277
Vermiform appendix, 269
Vertebrae
    cervical, 80
    lumbar, 83
    thoracic, 82
Vertebral artery, 229
Vertebral border of scapula, 87
Vertebral cavity, 164
Vertebral column, 80, 85
Vertebral foramen of thoracic vertebra, 82
Vertebral veins, 233
Vesicles, seminal, 285
Vessels
    blood, 239
        histology of, 239-242
    lymph, 251
Vestibular branch of vestibulocochlear nerve,
    203
Vestibule, 203
Vestibulocochlear nerve, 185, 203
Villi, 268
Visceral pericardium, 219
Visceral peritoneum, 269, 272
Visceral pleura, 256
Vision, 193-194
    binocular, and convergence, 200
    physiology of, 198-201
Visual acuity, Snellen chart, 199
Vital capacity, 263
Vitreous humor, 194
Volkmann's canals, 65
Volume, lung, 261-263
Voluntary muscle tissue, 129